Applications of Artificial Intelligence (AI) and Machine Learning (ML) in the Petroleum Industry

Today, raw data on any industry is widely available. With the help of artificial intelligence (AI) and machine learning (ML), this data can be used to gain meaningful insights. In addition, as data is the new raw material for today's world, AI and ML will be applied in every industrial sector. Industry 4.0 mainly focuses on the automation of things. From that perspective, the oil and gas industry is one of the largest industries in terms of economy and energy.

Applications of Artificial Intelligence (AI) and Machine Learning (ML) in the Petroleum Industry analyzes the use of AI and ML in the oil and gas industry across all three sectors, namely upstream, midstream, and downstream. It covers every aspect of the petroleum industry as related to the application of AI and ML, ranging from exploration, data management, extraction, processing, real-time data analysis, monitoring, cloud-based connectivity system, and conditions analysis, to the final delivery of the product to the end customer, while taking into account the incorporation of the safety measures for a better operation and the efficient and effective execution of operations.

This book explores the variety of applications that can be integrated to support the existing petroleum and adjacent sectors to solve industry problems. It will serve as a useful guide for professionals working in the petroleum industry, industrial engineers, AI and ML experts and researchers, as well as students.

Applications of Artificial Intelligence (AI) and Machine Learning (ML) in the Petroleum Industry

Manan Shah

Ameya Kshirsagar

Jainam Panchal

CRC Press
Taylor & Francis Group
Boca Raton London New York Leiden

CRC Press is an imprint of the
Taylor & Francis Group, an **informa** business

A BALKEMA BOOK

First published 2023
by CRC Press/Balkema

Schipholweg 107C, 2316 XC Leiden, The Netherlands
e-mail: enquiries@taylorandfrancis.com
www.routledge.com – www.taylorandfrancis.com

CRC Press/Balkema is an imprint of the Taylor & Francis Group, an informa business

© 2023 Manan Shah, Ameya Kshirsagar and Jainam Panchal

Library of Congress Cataloging-in-Publication Data
Names: Shah, Manan (Manan Rajiv), author. | Kshirsagar, Ameya,
author. | Panchal, Jainam, author.
Title: Applications of artificial intelligence (AI) and machine learning
(ML) in the petroleum industry / Dr. Manan Shah, Ameya Kshirsagar,
Jainam Panchal.
Description: First edition. | Boca Raton : CRC Press, 2023. | Includes
bibliographical references and index.
Identifiers: LCCN 2022008786 (print) | LCCN 2022008787 (ebook)
Subjects: LCSH: Petroleum engineering—Data processing. | Petroleum
industry and trade—Data processing. | Gas industry—Data processing.
Artificial intelligence—Industrial applications. |
Machine learning—Industrial applications.
Classification: LCC TN871 .S439 2023 (print) | LCC TN871 (ebook) |
DDC 338.2/7280285—dc23/eng/20220622
LC record available at https://lccn.loc.gov/2022008786
LC ebook record available at https://lccn.loc.gov/2022008787

ISBN: 978-1-032-24565-2 (hbk)
ISBN: 978-1-032-24588-1 (pbk)
ISBN: 978-1-003-27953-2 (ebk)

DOI: 10.1201/9781003279532

Typeset in Times New Roman
by codeMantra

Contents

Preface

Oil and gas are today's colossal pillars of the world energy sector. Things and resources we use in our daily life are connected directly or indirectly to the oil and gas industry. The oil and gas industry is essential to business perspectives as well for the nations' economic growth. The economic growth of any nation is related to its energy demand. As the energy demand is increasing and alternatives of it, i.e., renewable energy, is developing, oil and gas will be serving the world's need in the next upcoming years. Being one of the largest sectors in the world, it generates revenue of around US$3 trillion annually. If we see the Indian market, the import of crude oil elevates around US$102 billion in the financial years 2019–2020. This sector also incorporates a massive workforce globally; the generated revenues aid numerous households.

As the sector is immense and buttresses the economy, it requires more advancement in terms of digitalization, AI, and ML. The industrial revolution that is known as Industry 4.0 emphasizes the digitalization and automation of the industries and creates value products with the help of ML. Bringing the Internet of Things and cyber-physical systems into the sector, we can communicate, develop, forecast, and optimize things in real time.

The present book discusses about introducing AI and ML tools in the oil and gas industry to create value addition in the upstream, midstream, and downstream operations. From exploring to delivering to the appropriate customers, we can make the system digitalized and more efficient. There are several difficulties faced by the industry in transforming to digitalization and having a cloud-based system incorporated with the industrial internet of things which will provide real-time data and other vital aspects.

The variation in crude oil price depends on many factors, such as current demand, future demand, supply, and trading did over time. Based on the data for the past years and trends followed by it, and having the real-time data on hand, with the help of ML, models can be created for forecasting the prices considering all the influential parameters.

The present book will act as a guide for anyone who seeks to gain insights for bringing the methodologies: How to apply? Where to apply? When to apply and employ the various parameters of AI and ML for developing and understanding the technology?

The specific sections have been devoted to the applications of upstream, midstream, and downstream. ML has been discussed for accurate modeling, improvising subsurface characterizations, optimizing drilling operations, predictive maintenance, predicting operational outcomes, solving immediate problems, and replacing the manual workforce with robots at unsafe sites. Similarly, for AI, we have discussed the drilling operations, production, reservoir management, chatbots for the organizations providing services utilities, and monitoring the sites and automation. Moreover, incorporating different AI procedures can be applied effectively to tackle issues and establish technology advancements for better arrangements.

Moreover, ExxonMobil, a global leader in the oil and gas industry, has reached out to Massachusetts Institute of Technology to structure and design the AI robots for sea investigation aiming to improve the normal oil leak identification abilities. The present book will motivate the developers with the students and universities to develop and design such extraordinary systems for enhancing the oil and gas operation.

This book is a ready reference for curious readers and keen learners who wish to join the new revolution and contribute to this field. We believe that AI and ML can assuage risk, enrich productivity, and minimize operational costs. It has precise targeting and ability to identify appropriate drilling locations, which will in turn maximize Return On Investment (ROI) on the related activities. The demand for AI in the upstream market is set to accelerate at the rate of 12.66% to reach around $3 billion by 2022–2023.

About the authors

Manan Shah is a B.E. in chemical engineering from LD College of Engineering and an M.Tech. in petroleum engineering from the School of Petroleum Technology, Pandit Deendayal Petroleum University (PDPU). He has completed his Ph.D. in the area of exploration and exploitation of geothermal energy in the state of Gujarat. He is currently working as an Assistant Professor in the Department of Chemical Engineering, School of Technology, PDPU, and he has published several articles in reputed international journals in the area of many sectors. He serves as an active reviewer for several reputable international journals like Springer and Elsevier. Dr. Shah is a young and dynamic academician and researcher in various engineering fields. Dr. Shah has received the best paper award for the world's second-best journal.

Ameya Kshirsagar is B.Tech. in information technology from Symbiosis Institute of Technology with gusto to discover value-added aspects of computer science, data analytics and data science, and competency to take on every learning challenge. He has published several journal articles in reputed international journals in computer science, data science, ML, finance, healthcare, etc. Besides this, he also contributed to a book published by IOP by writing a comprehensive chapter on multimedia security. He is currently targeting foreign institutes for his higher education in computer science and data science.

Jainam Panchal is currently pursuing his B.Tech. in the chemical engineering discipline from the Pandit Deendayal Energy University, Gandhinagar, Gujarat. He has done his schooling at Best Higher Secondary School, Maninagar, Ahmedabad. He is an enthusiastic, creative, and ethical technofreak on the journey of learning following his passion with joy. He is interested in the energy sector and renewable energy domain with a core chemical domain. He has been an active participant in the core-curricular activities not only in his school days but also now. He has published many good articles in reputed journals. He is strong in his leadership skills and interpersonal skills. He also focuses more on teamwork and time management for an effective and efficient work output. His hobbies include traveling and exploring, cinema, and food.

Chapter 1

A comprehensive review of machine application in the oil and gas industry

1.1 INTRODUCTION

Oil and gas are one of the most crucial artifacts (Longwell, 2002). When converted, it can be used to power vehicles, asphalt, electricity, and heat generation (Asche et al., 2006; Oliveira-Pinto et al., 2019). Furthermore, petroleum is used latently in a variety of products such as plastics, paints, chemicals, tape, clothes, toothbrushes, football, combs, CDs and DVDs, paintbrushes, detergents, vaporizers, balloons, sunglasses, stents, heart valves, crayons, parachutes, telephone enamel, cameras, anesthetics, artificial turf, artificial limbs, bandages, dentures, cold cream, movie film, and soft contact lenses (Coady et al., 2007; Chisholm, 2015). It is difficult to imagine a world without oil and gas.

According to IBIS World, the global oil and gas exploration and production industry is worth $3.3 trillion in 2021 (IBISWorld, 2020). The oil and gas industry, when combined with machine learning, can produce more extraordinary isolated facility handling and can also act in real-time for easing the process, relieving the manual workforce, easing land analysis, safety, and regulatory issues, and so on (Anifowose et al., 2019).

Machine learning and advanced technologies are the new imminent strategy to alleviate the oil and gas industry issues and add business value (Al-Jamimi et al., 2018). These sectors and firms aspire to take up the novel machine learning technology, but they struggle for the same due to lack of machine learning knowledge (Qi et al., 2018), which results in the inability to provide efficient outcomes from the models, plausible results or returns, and investments (Pirizadeh et al., 2021).

Figure 1.1 shows that oil and gas are majorly divided into the following three sections (Kalita, 2020):

- **Upstream:** Here, the main job is to explore, i.e., drilling to find oil and extracting it; in simple terms, finding spots, production, and processing.
- **Midstream:** This stage comprises transportation and storage of oil and gas.

Downstream: This is either aimed at the end user or is further refined before being sent to marketing firms.

- We can divide machine learning in the oil and gas industry into three categories: experimentation, operationalization, and maintenance.

DOI: 10.1201/9781003279532-1

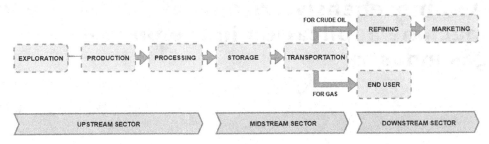

Figure 1.1 Steps followed in the oil and gas industry.

- Experimentation encompasses data collection and processing, algorithm selection validation, and verification. Besides, operationalization focuses on deploying models and monitoring quantifiable services (Lehr and Ohm, 2017).

So far, the application of machine learning in the oil and gas industry has been limited to using deep learning methodologies for prediction and seismic analysis (Li et al., 2018; Wrona et al., 2018; Kiani et al., 2019). However, due to availability of a large amount of overflowing data generated by the industry, several applications are rising simultaneously. This collection, analysis, and data interpretation can further yield fruitful results if properly bridged with machine learning techniques.

Machine learning techniques can further enhance the capability of the oil and gas sector by optimizing extraction, streamlining the workforce, and delivering accurate models (Kamal, 2020). Moreover, some of the ways in which machine learning can aid in the oil and gas sector are accurate modeling, analysis and decision-making of digging sites, analysis and enhancement of subsurface characterization, optimization of drilling procedures (Hegde and Gray, 2017, 2018; Ma et al., 2019), prediction and delay of maintenance, prediction of price, etc.

In the following sections, we will highlight some of the applications of machine learning in the oil and gas industry and how they mitigate risk—followed by a brief examination of machine learning applications in upstream, midstream, and downstream operations.

1.2   CONTRIBUTION OF MACHINE LEARNING TO THE OIL AND GAS INDUSTRY

Though there is slow progress in the oil and gas industry field with respect to machine learning application, there is evidence of some machine learning work done in this field in the oil and gas industry to assuage the issues and improve overall efficiency. In this section, we will briefly look into some machine learning applications in the oil and gas industry.

1.2.1   Predictive maintenance

Predicting the occurrence of maintenance can increase productivity tenfold by preventing all types of mishaps caused by absenteeism in maintenance work due to human error

(Susto et al., 2015). It will also result in an efficient system resulting in a stable company. Predictive maintenance can be used to keep a watch in real-time on the tanks' level and volumes, and pipeline quality analysis to gauge the imminent required maintenance work (Carvalho et al., 2019). Moreover, apart from this daily throughput to be kept in check, we will analyze and predict how atmospheric conditions can affect the types and pieces of equipment, contents, business, and user demands (Aissani et al., 2009).

As shown in Figure 1.2, one of the best examples of predictive maintenance with the use of machine learning is done by General Electric Company, "Predix" (Erickson et al., 2015; Immelt, 2017), which aided them in reducing employee routine time to process by 60% and fuel usage by 20%. Additionally, due to predictive maintenance implementation, Predix (Lohr, 2106), the equipment has shown a 5% increase in uptime. Predix is a platform incorporated by General Electric Company with the use of the Internet of Things for a major collection of data (Kanawaday and Sane, 2018). It employs machine learning techniques to aid industrial systems to assuage it.

1.2.2   Spotting digging sites with machine learning

Digging sites refers to the locations where wells can be dug to extract oil or gas. Conventionally these wells were spotted using a reservoir simulator as a function evaluator (Nwachukwu et al., 2018). Nevertheless, this model was highly time-consuming, expensive, and inefficient based on the efforts applied. However, this inefficiency can be overcome using machine learning to pinpoint the digging sites or wells in this specific sector (Nasir et al., 2020). Several aspects are taken into consideration when a well is dug (Leem et al., 2015); this data can be further used for spatial analysis (Zhang et al., 2016). This methodology is primarily supported by the analysis of intricate petrophysical and geological models by interpreting the seismic data (Maxwell et al., 2010; Chu et al., 2020). Due to this, one can get a comprehensive picture of the geology of the area. With the analysis of this combined information, oil and gas wells were pinpointed under the rock (Zhang et al., 2016). Due to the large quantity of data, the model can be more accurate, fast, and reliable (Sharifi et al., 2014). This will save time by reducing the number of boreholes and avoiding the need to dig test holes, as well as money.

Moreover, this land mapping (Naghibi et al., 2016) can further be made portable for ease of use and making the process more efficient.

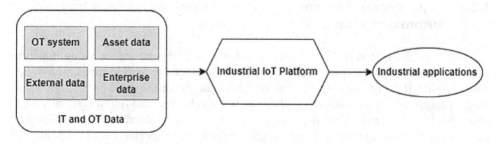

Figure 1.2 Predix platform.

1.2.3   Machine learning in drilling operations

Drilling in the oil and gas industry is referred to as creating a well approximately 12–100 cm wide in diameter into the earth with the help of a drilling rig followed by placement of a steel pipe casing (Ma et al., 2016). This process entails a large number of expert people working together to complete this complex task in a timely manner in order to ensure smoother operation, safety, and accuracy.

Due to conventional methodologies currently, the oil and gas industry uses physics-based models to tackle the issues. These issues comprise stick-slip vibration/hole cleaning, pipe failures, loss of circulation, Basics and Corrective Actions (BHA) whirl, stuck pipe incidents, excessive torque and drag, low Rate of Penetration (ROP), bit wear, formation damage, and borehole instability (Noshi and Schubert, 2018). These problems can be mitigated to a greater extent with the use of machine learning techniques. Modern rigs are installed with several sensors to collect relevant number real-time metadata and data, which can further be coupled with machine learning to infer advanced computer vision-based video information (Hegde and Gray, 2017), saving time and manual workforce with maximum possible accuracy (Zainal Abidin et al., 2019). Though machine learning and data mining techniques have several applications in the industry, the oil and gas industry is yet to venture them to the fullest (Zhao et al., 2020). Royal Dutch Shell has implemented AI and machine learning-driven assistants (Emma and Ethan) (specifically natural language processing) to suggest and advise end-to-end coupled drilling optimization, lubricant, and related products (Marr, 2020).

1.2.4   Problem-solving with machine learning

As we have discussed, machine learning can mitigate the issues faced by the oil and gas industry. It can further help in solving complex problems with the utmost possible efficiency. Machine learning can be used as a case-based reasoning tool. Given the vast tracks and records of problems encountered in the past, machine learning can quickly sift through the massive dataset to point out the most similar past cases in the massive dataset and provide the same solutions (Sun et al., 2016). The overall process of collecting data and comparing it to previous data with machine learning techniques to provide solutions can save time, money, and labor.

1.2.5   Replacement of manual labor with machine learning tools or automated robots

Furthermore, a portion of the manual workforce can be replaced by using machine learning tools and techniques. It will not entirely replace the workforce as human operatives will still be required to manage the tasks and make the throughput higher and more efficient (Sun and Yan, 2019). These skilled workers would require a hybrid skill and knowledge of geoscience, mathematics, coding, and problem-solving. These machine learning techniques can help enhance the customer experience and keep the cost at par in the oil and gas industry (Mcafee and Brynjolfsson, 2012).

In this section, we shed some light on how machine learning can be inculcated in the oil and gas industry and how it can mitigate the problems, reduce the time and expenses, and improve the overall efficiency of the holistic process. In the next section, we will discuss how machine learning can be incorporated into different oil and gas industry sectors, i.e., the upstream, midstream, and downstream.

1.3   MACHINE LEARNING IN THE OIL AND GAS UPSTREAM SECTOR

In Figure 1.3, we can see that the upstream sector majorly consists of exploration, production, and processing of oil and gas. Here exploration and productions are the early steps taken in the production of the final good (Chen, 2020). The workers explore, discover, and dig wells to extract oil and gas from below the earthen surface in exploration (Kraus, 2020). Moreover, production is the process of bringing up crude oil and gas to the surface. In the last phase of the upstream sector, "processing," oil and gas are worked upon to separate, discard, or convert the mixture into different components (Soam, 2019). This mixture may consist of several liquid hydrocarbons, gas, water, and solids. After separation, these goods move toward the midstream sector. Here we will look into some applications of machine learning in the upstream sector.

Big data and machine learning can contribute to exploration and production of the oil and gas industry. It can significantly reduce the need for excessive materials, workforce, and needless logistics. Feblowitz et al. (2013) stated that big data machine learning techniques could be utilized in the form of pattern recognition. With the robust dataset of seismic data, it can point out the productive seismic trace signatures. Furthermore, Microsoft (2015) applications performed on the basis of seismic trace pattern recognition (Wang et al., 2018). Furthermore, according to Baaziz and Quoniam (2019), there is a greater need for seismic data, its geometry, and comparison with neighboring wells in order to analyze the 3D pattern and detect hydrocarbon deposits.

As suggested above, one can analyze the pattern and spot the wells. ESDS (2016) stated that data collected on weather, soil, and sensor can be utilized to discover new oil deposits accurately, ease the process of drilling operations and reduce the overall cost (Anifowose et al., 2016).

Production is another crucial step in the upstream oil and gas industry, and its forecasting can play a vital role in project planning and can influence economic development (Maitland, 2000). Moreover, it will also aid in the further process of installation of artificial lifts, type of workforce, operations, and facilities design.

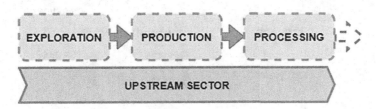

Figure 1.3 Upstream sector of the oil and gas industry.

Li et al. (2013) presented oil production predictions with a neural decision tree model parallel to C4.5 and Artificial Neural Network (ANN). Considering Mean Absolute Error (MAE) and Mean Squared Error (MSE) as metrics of measurement, the neural decision tree model has the least error value proving to be better than C4.5 and ANN. On the other hand, Chakra et al. (2013) practiced a higher order neural network to predict the oil production rate. The model was quite satisfactory in measuring its performance on the metrics of MSE, RMSE, and mean absolute percentage error. However, there was a requirement of tweaking with respect to the well's nature, i.e., static and dynamic reservoir. Choubineh et al. (2017) conducted the same experiment with hybrid ANN with six most influential attributes, oil specific gravity, gas specific gravity, gas–liquid ratio, choke size, wellhead pressure, and temperature. On the contrary Liu et al. (2020) used hybrid models with ensemble empirical mode decomposition (EEMD)-based LSTM models that yielded the lowest MAE value when compared to EEMD-ANN and EEMD-SVM.

Once the oil and gas spots are explored, located, and brought up to the surface (production), they move on to the next phase, i.e., the midstream sector.

1.4 MACHINE LEARNING IN THE OIL AND GAS MIDSTREAM SECTOR

In Figure 1.4, we can see that the midstream sector majorly consists of storage and transportation of oil and gas. The midstream oil and gas industry sector mostly consists of storage, processing, and transportation of crude oil and gas products (White et al., 2018). The storage part majorly consists of facilities or premises for storage (Hsu and Robinson, 2019). These include bulk terminals, refinery tanks, and holding tanks. On the other hand, in terms of transportation, oil and gas are primarily transported from one location to another using two methods (Hsu and Robinson (2019): first, through tankers running on water, and second through pipelines that may or may not run underground (Hsu and Robinson, 2019). Moreover, the monitoring of these stages and their maintenance is also a part of the midstream sector (Nadj et al., 2016). Furthermore, we will see how machine learning techniques can be used in the midstream sector.

Transferring unknown quality, volumes, and grades of goods (oil, gas, and other forms) from multiple locations to customers and markets is a challenging task to achieve. With the help of machine learning, several issues of the midstream sector can be mitigated. Machine learning analytics can be used for the connected pipelines, leak detection sensors, alarms, and several other measurable and influential quantities with the sensors (Dixit et al., 2018). This sector majorly benefits from big data analytics, part

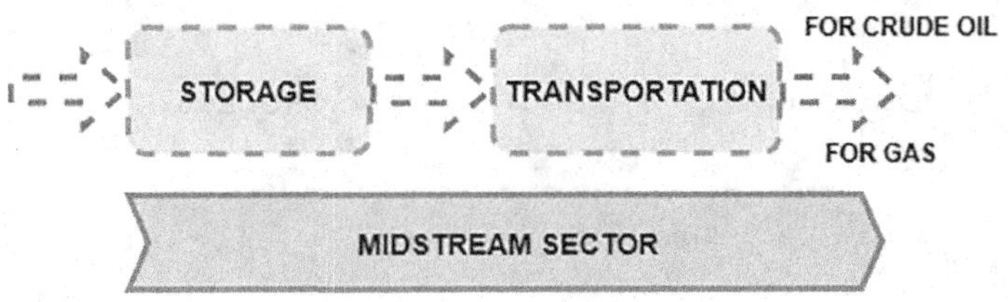

Figure 1.4 Midstream sector of the oil and gas industry.

of machine learning, for the case of pipeline operations, the environmental monitoring and infrastructure management can be done using the SCADA application (Kivi, 2017).

On the other hand, Kohlleffel (2015) focused on how Hadoop is deployed on data-enabled models to use machine learning techniques to provide a view, process, and deliver predictive analysis. Big data analytics is also utilized in the transport and refinement of oil and gas (Dixit et al., 2018). Because of the novel sensory tool, data collection, data logging, and examination are relatively more straightforward comparatively. This information and analysis can be extrapolated to predict the imminent equipment examination and maintenance.

CAR (2018) emphasized that DCP Midstream, a firm consisting of over 50,000 miles of natural gas pipeline and more than 60 processing plants, is a significant contributor to the midstream segment of the oil and gas industry (Rignall, 2018). DCP had an extensive collection of sensory collected data. Furthermore, they aggregated this data with machine learning techniques to bear fruitful results. They analyzed and closely looked at the specific equipment performance and maintenance. The information gained (patterns) from sensors is fed to machine learning to detect the emergence of attention a machine requires before it fails (Grainger Editorial Staff, 2018). By employing this technology, DCP Midstream has significantly improved compressor reliability and expenses on equipment.

After storing and transporting the extracted product, the oil and gas industry's main target is to either refine the crude product or sell it to the market altogether.

1.5   MACHINE LEARNING IN THE OIL AND GAS DOWNSTREAM SECTOR

As shown in Figure 1.5, the downstream sector of the oil and gas industry majorly consists of refining and distribution of these goods. The refining part is responsible for converting oil and gas into the final forms such as gasoline, natural gas liquids,

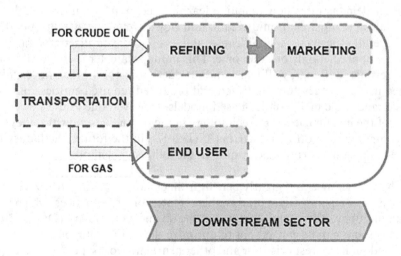

Figure 1.5 Downstream sector of the oil and gas industry.

polymer, plastics, lubricants, waxes, diesel, and several energy sources (Risse, 2019). In the distribution part, the final goods are distributed among the market, wholesalers, retail, customers or the end-users, and their service, as distribution is the final step in the oil and gas industry (Davcheva, 2019).

Azadeh et al. (2008) focused majorly on two models in oil and gas refinery for energy analysis. He used data envelopment analysis and principal component analysis. First, data envelopment analysis models are majorly used for estimating the efficiency in a given data set for decision-making units. Azadeh et al. (2008) made use of input-oriented data envelopment analysis models to evaluate the efficaciousness of refineries under a fixed structure. Furthermore, based on the scores received by the input-oriented data envelopment analysis model, they retrospected the production of refinery products with the least expense of electricity and fuels.

On the other hand, Tokarek et al. (2018) used principal component analysis to gauge the pollutants present in the Athabasca oil sands. On analysis, they rectified 28 most influential variables, among which volatile organic compounds and intermediate-volatility organic compounds are the most influential.

Fan et al. (2019) emphasized the distribution pipeline's efficiency, which is based on the delta of energy and volume of the content. The authors used a data envelope analysis model to study the delta change in the input and output energy and volume, thereby ascertaining the pipeline's efficiency.

As in Figure 1.6, Yu et al. (2019) focused on the oil and gas industry's market trend by researching the various factors and trends affecting oil and gas consumption. Yu et al. (2019) utilized Google trends to study future oil consumption through a big data and machine learning-driven prediction model. The model works in two parts: first being investigation employed via co-integration test and Granger analysis, and second being prediction improvement by implementing Google trends to both the models (statistical and artificial intelligence model).

1.6 CHALLENGES AND FUTURE SCOPE

The chapter demonstrates how machine learning techniques can be used in the oil and gas industry's upstream, midstream, and downstream petroleum operations to improve overall efficiency. A significant number of studies utilized several parameters to yield or predict only one outcome. This input parameter is collected from core analysis, seismic, well logs, well tests, and production history. Machine learning-implemented models successfully analyzed and predicted required entities compared to the physics and basic mathematics-based models. However, due to the slower acceptance rate of the machine learning technology in the oil and gas industry, they still lag behind compared with other industries (Kivi, 2017). In the future, the industries can make use of machine learning techniques in the following methods:

- Machine learning could greatly benefit from cloud computing. It would help perform more complex analyses and enable remote working practices, which will also result in safer working environments in the oil and gas industry (Dyson, 2017).
- Investigations can be carried out to optimize the drilling attributes using models like random K-nearest neighbor and other neural networks' performances (Hegde et al., 2020).

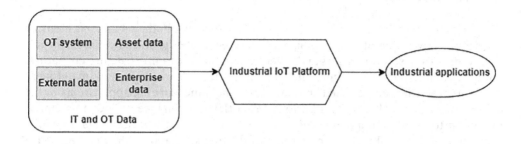

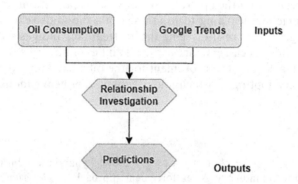

Figure 1.6 Framework of big data-driven oil consumption prediction model employing Google trends.

- Though Support Vector Machine (SVM) has yielded significant accuracy to spot drilling locations, more efficient models like ELM can further increase the accuracy (Qian et al., 2018).
- For prediction of complex reservoirs, a more extensive data set should be used to avoid underfitting and yield accurate results (Aulia et al., 2010).
- Due to the absence of logs of the complete wellbore interval, there has not been much study on this aspect; hence, more data can be collected adequately to propose better working methodologies (Gholami et al., 2014).
- Careful sections of the attribute with a correlation matrix and normalization technique can result in better-performing prediction models (Salehi et al., 2017).
- Where machine learning models fail to reach the requirements, deep learning models can be tested to check for a higher outcome rate.
- Regularization can help in overcoming the underfit and overfit problems.
- Advanced sensors can be used for collection of accurate data.

1.7 CONCLUSION

This chapter discussed the application of machine learning in the oil and gas industry and showed how different methodologies could affect it. As the oil and gas industry collects its data from several sensors and other devices that are multi-structured in nature, novel tools should be used to integrate and analyze them to yield meaningful outputs.

Moreover, this chapter sheds light on some machine learning applications in the oil and gas industry. It can be used to forecast the maintenance periods to enhance the life of tools and equipment, reducing the expenses and workload. Additionally, machine learning applications in spotting of digging sites and drilling can also save much time, and skilled manual labor and machine learning models can be used to extrapolate the prior scenarios to mend the currently existing issue of similar nature.

Then, in brief, we discussed the three different machine learning use cases in oil and gas. i.e., upstream, midstream, and downstream. We majorly focused on machine learning applications in exploration, production, processing, storage, transportation, refining, and marketing of the end product.

We observed that the upstream sector has a greater number of existing implementations of machine learning techniques, while the midstream and downstream sectors are lagging. As a result, seeing the upstream oil and gas sector adapt to machine learning techniques can encourage the oil and gas industry as a whole to do the same. There is still a scope of improvement for the oil and gas industry as they are lagging and slower in adopting new technologies than other heavy industries.

REFERENCES

Aissani, N., Beldjilali, B., & Trentesaux, D. (2009). Dynamic scheduling of maintenance tasks in the petroleum industry: A reinforcement approach. *Engineering Applications of Artificial Intelligence, 22*, 1089–1103. https://doi.org/10.1016/j.engappai.2009.01.014

Al-Jamimi, H. A., Al-Azani, S., & Saleh, T. A. (2018). Supervised machine learning techniques in the desulfurization of oil products for environmental protection: A review. *Process Safety and Environmental Protection, 120*, 57–71. https://doi.org/10.1016/j.psep.2018.08.021

Anifowose, F., Abdulraheem, A., & Al-Shuhail, A. (2019). A parametric study of machine learning techniques in petroleum reservoir permeability prediction by integrating seismic attributes and wireline data. *Journal of Petroleum Science and Engineering, 176*, 762–774. https://doi.org/10.1016/j.petrol.2019.01.110

Anifowose, F., Adeniye, S., Abdulraheem, A., & Al-Shuhail, A. (2016). Integrating seismic and log data for improved petroleum reservoir properties estimation using non-linear feature-selection based hybrid computational intelligence models. *Journal of Petroleum Science and Engineering, 145*, 230–237. https://doi.org/10.1016/j.petrol.2016.05.019

Asche, F., Osmundsen, P., & Sandsmark, M. (2006). The UK market for natural gas, oil and electricity: Are the prices decoupled? *The Energy Journal, 27*, 27–40. https://doi.org/10.5547/ISSN0195-6574-EJ-Vol27-No2-2

Aulia, A., Keat, T. B., Maulut, M. S., et al. (2010). Smart oilfield data mining for reservoir analysis. *International Journal of Engineering & Technology, 10*, 78–88.

Azadeh, A., Ghaderi, S. F., & Asadzadeh, S. M. (2008). Energy efficiency modeling and estimation in petroleum refining industry - A comparison using physical data. *Renewable Energy Power Quality Journal, 1*, 123–128. https://doi.org/10.24084/repqj06.242

Baaziz, A., & Quoniam, L. (2019). How to use big data technologies to optimize operations in upstream petroleum industry. *International Journal of Innovation*. https://doi.org/10.2139/ssrn.3429410

CAR. (2018). Using machine learning to improve how energy gets to market - technology and operations management. In: HBS Digit. Initiat. https://digital.hbs.edu/platform-rctom/submission/using-machine-learning-to-improve-how-energy-gets-to-market/. Accessed 22 Jan 2021.

Carvalho, T. P., Soares, F. A., Vita, R., et al. (2019). A systematic literature review of machine learning methods applied to predictive maintenance. *Computers & Industrial Engineering*, *137*, 106024. https://doi.org/10.1016/j.cie.2019.106024

Chakra, N. C., Song, K. Y., Gupta, M. M., & Saraf, D. N. (2013). An innovative neural forecast of cumulative oil production from a petroleum reservoir employing higher-order neural networks (HONNs). *Journal of Petroleum Science and Engineering*, *106*, 18–33. https://doi.org/10.1016/j.petrol.2013.03.004

Chen, J. (2020). Exploration & production (E&P). In: Investopedia. https://www.investopedia.com/terms/e/exploration-production-company.asp. Accessed 21 Jan 2021.

Chisholm, K. (2015). 144 products made from petroleum and some that may may shock you. In: IAG wealth Manag. https://innovativewealth.com/inflation-monitor/what-products-made-from-petroleum-outside-of-gasoline/. Accessed 23 Jan 2021.

Choubineh, A., Ghorbani, H., Wood, D. A., et al. (2017). Improved predictions of wellhead choke liquid critical-flow rates: Modelling based on hybrid neural network training learning based optimization. *Fuel*, *207*, 547–560. https://doi.org/10.1016/j.fuel.2017.06.131

Chu, M., Min, B., Kwon, S., et al. (2020). Determination of an infill well placement using a data-driven multi-modal convolutional neural network. *Journal of Petroleum Science and Engineering*, *195*, 106805. https://doi.org/10.1016/j.petrol.2019.106805

Coady, D., Baig, T., Ntamatungiro, J., & Mati, A. (2007). Domestic petroleum product prices and subsidies: Recent developments and reform strategies. IMF Work Pap 07:1. https://doi.org/10.5089/9781451866353.001

Davcheva, M. (2019). Oil and gas industry overview. In: ScheduleReader. https://www.schedulereader.com/blog/oil-and-gas-industry-overview. Accessed 23 Jan 2021.

Dixit, G., Blaney, J., Dixit, H., & Michael, A. (2018). TWA article blockchain technology for the oil and gas industry. https://pubs.spe.org/en/twa/twa-article-detail/?art=4265. Accessed 28 Oct 2020.

Dyson, R. (2017). The oil and gas industry needs to think big with Big Data. In: Powerful Think. http://ioconsulting.com/offshore-engineer-big-data-interview/. Accessed 23 Jan 2021.

Erickson, J., Klasky, H., & Beyah-Taylor, C. (2015). *Creation of GE Digital | Business Wire*. Businesswire. https://www.businesswire.com/news/home/20150914006029/en/Creation-GE-Digital. Accessed 20 Jan 2021.

ESDS. (2016). Big value for big data in oil and gas industry! In: ESDS. https://www.esds.co.in/blog/big-value-big-data-oil-gas-industry/#sthash.eRZ9Zjt0.FaWDdTJz.dpbs. Accessed 21 Jan 2021.

Fan, M., Ao, C., & Wang, X. (2019). Comprehensive method of natural gas pipeline efficiency evaluation based on energy and big data analysis. *Energy 188*, 116069. https://doi.org/10.1016/j.energy.2019.116069

Feblowitz, J., Rice, L., Beals, B., et al. (2013). *Big Data in Oil and Gas: How to Tap its Full Potential - PDF Free Download*. Hitachi Data Syst. http://docplayer.net/1497544-Big-data-in-oil-and-gas-how-to-tap-its-full-potential.html. Accessed 21 Jan 2021.

Gholami, R., Moradzadeh, A., Maleki, S., et al. (2014). Applications of artificial intelligence methods in prediction of permeability in hydrocarbon reservoirs. *Journal of Petroleum Science and Engineering*, *122*, 643–656. https://doi.org/10.1016/j.petrol.2014.09.007

Grainger Editorial Staff. (2018). Predictive maintenance programs & AI. In: Grainger Kno-wHow. https://www.grainger.com/know-how/industry/manufacturing/kh-predictive-mainte-nance-ai-preventativemaintenance-strategy. Accessed 22 Jan 2021.

Hegde, C., & Gray, K. (2018). Evaluation of coupled machine learning models for drilling optimization. *Journal of Natural Gas Science and Engineering, 56*, 397–407. https://doi.org/10.1016/j.jngse.2018.06.006

Hegde, C., & Gray, K. E. (2017). Use of machine learning and data analytics to increase drilling efficiency for nearby wells. *Journal of Natural Gas Science and Engineering, 40*, 327–335. https://doi.org/10.1016/j.jngse.2017.02.019

Hegde, C., Pyrcz, M., Millwater, H., et al. (2020). Fully coupled end-to-end drilling optimization model using machine learning. *Journal of Petroleum Science and Engineering, 186*, 106681. https://doi.org/10.1016/j.petrol.2019.106681

Hsu, C. S., & Robinson, P. R. (2019). Midstream transportation, storage, and processing. In: *Petroleum Science and Technology.* Springer International Publishing, pp. 385–394.

IBISWorld. (2020). *Global Oil & Gas Exploration & Production – Market Size.* IBISWorld. https://www.ibisworld.com/global/market-size/global-oil-gas-exploration-production/. Accessed 23 Jan 2021.

Immelt, J. R. (2017). *How I Remade GE. Harv. Bus. Rev.* https://hbr.org/2017/09/how-i-re-made-ge. Accessed 20 Jan 2021.

Kalita, I. (2020). The oil and gas industry of Assam- The upstream, downstream and midstream industry. *Palarch's Journal of Archaeology of Egypt/Egyptology, 17*, 13252–13267.

Kamal, M. M. (2020). Future need of petroleum engineering. *SPE West Reg Meet Proc 2020-April*, 1–14. https://doi.org/10.2118/200771-ms

Kanawaday, A., & Sane, A. (2018). Machine learning for predictive maintenance of industrial machines using IoT sensor data. In: *Proceedings of the IEEE International Conference on Software Engineering and Service Sciences, ICSESS.* IEEE Computer Society, pp. 87–90.

Kiani, J., Camp, C., & Pezeshk, S. (2019). On the application of machine learning techniques to derive seismic fragility curves. *Computers & Structures, 218*, 108–122. https://doi.org/10.1016/j.compstruc.2019.03.004

Kivi, J. (2017). *How the Internet of Things Has Influenced Midstream Pipeline Operations.* In: Schneider Electr. Blog. https://blog.se.com/industrial-software/2017/10/03/iiot-midstream-pipe-line-operations/. Accessed 23 Jan 2021.

Kohlleffel, K. (2015). *A Petrophysicist's Perspective on Hadoop-based Data Discovery.* In LinkedIn. https://www.linkedin.com/pulse/petrophysicists-perspective-hadoop-based-data-kel-ly-kohlleffel/. Accessed 23 Jan 2021.

Kraus, R. S. (2020). Chapter 75- Oil exploration and drilling. In: *Encycl. Occup. Heal. Saf.* http://www.ilocis.org/documents/chpt75e.htm. Accessed 21 Jan 2021.

Leem, J., Lee, K., Kang, J. M., et al. (2015). History matching with ensemble Kalman filter using fast marching method in shale gas reservoir. *Soc Pet Eng–SPE/IATMI Asia Pacific Oil Gas Conf Exhib APOGCE 2015*, 1–19. https://doi.org/10.2118/176164-ms

Lehr, D., & Ohm, P. (2017). Playing with the data: What legal scholars should learn about machine learning. *UC Davis Law Rev, 51*, 653–717.

Li, W., Narvekar, N., Nakshatra, N., et al. (2018). Seismic data classification using machine learning. *Proc - IEEE 4th Int Conf Big Data Comput Serv Appl BigDataService 2018*, 56–63. https://doi.org/10.1109/BigDataService.2018.00017

Li, X., Chan, C. W., & Nguyen, H. H. (2013). Application of the Neural Decision Tree approach for prediction of petroleum production. *Journal of Petroleum Science and Engineering, 104*, 11–16. https://doi.org/10.1016/j.petrol.2013.03.018

Liu, W., Liu, W. D., & Gu, J. (2020). Forecasting oil production using ensemble empirical model decomposition based long short-term memory neural network. *Journal of Petroleum Science and Engineering, 189*, 107013. https://doi.org/10.1016/j.petrol.2020.107013

Lohr, S. (2016). G.E., the 124-Year-Old Software Start-Up. *New York Times.* https://www.nytimes.com/2016/08/28/technology/ge-the-124-year-old-software-start-up.html?_r=0. Accessed 20 Jan 2021.

Longwell, H. J. (2002). The future of the oil and gas industry: Past approaches, new challenges. *World Energy, 5,* 100–104.

Ma, T., Chen, P., & Zhao, J. (2016). Overview on vertical and directional drilling technologies for the exploration and exploitation of deep petroleum resources. *Geomech Geophys Geo-Energy Geo-Resources, 2,* 365–395. https://doi.org/10.1007/s40948-016-0038-y

Ma, Z., Vajargah, A. K., Lee, H., et al. (2019). Applications of machine learning and data mining in Speedwise® drilling analytics: A case study. *Paper presented at the Abu Dhabi International Petroleum Exhibition & Conference 2018, ADIPEC 2018.* https://doi.org/10.2118/193224-ms

Maitland, G. C. (2000). Oil and gas production. *Current Opinion in Colloid & Interface Science, 5,* 301–311. https://doi.org/10.1016/S1359-0294(00)00069-8

Marr, M. (2020). The incredible ways shell uses artificial intelligence to help transform the oil and gas giant. In: Bernard Marr Co. https://bernardmarr.com/default.asp?contentID=1796. Accessed 21 Jan 2021.

Maxwell, S. C., Rutledge, J., Jones, R., & Fehler, M. (2010). Petroleum reservoir characterization using downhole microseismic monitoring. *Geophysics, 75.* https://doi.org/10.1190/1.3477966

Mcafee, A., & Brynjolfsson, E. (2012). Spotlight on Big Data: Big Data: The management revolution. *Harvard Business Review, 1–9.*

Microsoft. (2015). Drilling for new business value: How innovative oil and gas companies are using big data to outmaneuver the competition. https://cloudblogs.microsoft.com/industry-blog/uncategorized/2015/10/18/drilling-new-business-value-innovative-oil-gas-companies-using-big-data-maneuver-competition/. Accessed 21 Jan 2021.

Nadj, M., Jegadeesan, H., Maedche, A., et al. (2016). *A Situation Awareness Driven Design for Predictive Maintenance Systems: The Case of Oil and Gas Pipeline Operations.* United States: The Association for Information Systems (AIS).

Naghibi, S. A., Pourghasemi, H. R., & Dixon, B. (2016). GIS-based groundwater potential mapping using boosted regression tree, classification and regression tree, and random forest machine learning models in Iran. *Environmental Monitoring and Assessment, 188,* 1–27. https://doi.org/10.1007/s10661-015-5049-6

Nasir, Y., Yu, W., & Sepehrnoori, K. (2020). Hybrid derivative-free technique and effective machine learning surrogate for nonlinear constrained well placement and production optimization. *Journal of Petroleum Science and Engineering, 186,* 106726. https://doi.org/10.1016/j.petrol.2019.106726

Noshi, C. I., & Schubert, J. J. (2018). The role of machine learning in drilling operations; a review. In: *SPE Eastern Regional Meeting, Pittsburgh, Pennsylvania, USA, October 2018.* Paper Number: SPE-191823-18ERM-MS. One Petro – Society of Petroleum Engineers (SPE).

Nwachukwu, A., Jeong, H., Pyrcz, M., & Lake, L. W. (2018). Fast evaluation of well placements in heterogeneous reservoir models using machine learning. *Journal of Petroleum Science and Engineering, 163,* 463–475. https://doi.org/10.1016/j.petrol.2018.01.019

Oliveira-Pinto, S., Rosa-Santos, P., & Taveira-Pinto, F. (2019). Electricity supply to offshore oil and gas platforms from renewable ocean wave energy: Overview and case study analysis. *Energy Conversion and Management, 186,* 556–569. https://doi.org/10.1016/j.enconman.2019.02.050

Pirizadeh, M., Alemohammad, N., Manthouri, M., & Pirizadeh, M. (2021). A new machine learning ensemble model for class imbalance problem of screening enhanced oil recovery methods. *Journal of Petroleum Science and Engineering, 198,* 108214. https://doi.org/10.1016/j.petrol.2020.108214

Qi, G., Zhu, Z., Erqinhu, K., et al. (2018). Fault-diagnosis for reciprocating compressors using big data and machine learning. *Simulation Modelling Practice and Theory, 80,* 104–127. https://doi.org/10.1016/j.simpat.2017.10.005

Qian, K. R., He, Z. L., Liu, X. W., & Chen, Y. Q. (2018). Intelligent prediction and integral analysis of shale oil and gas sweet spots. *Petroleum Science*, *15*, 744–755. https://doi.org/10.1007/s12182-018-0261-y

Rignall, T. (2018). *Enabling Business Transformation with the PI System: The DCP 2.0 Journey*. U.S.A.: OSI World.

Risse, M. (2019). Refineries need machine learning to improve operations. In: Oil Gas Eng. https://www.oilandgaseng.com/articles/refineries-need-machine-learning-to-improve-operations/. Accessed 23 Jan 2021.

Salehi, M. M., Rahmati, M., Karimnezhad, M., & Omidvar, P. (2017). Estimation of the non records logs from existing logs using artificial neural networks. *Egyptian Journal of Petroleum*, *26*, 957–968. https://doi.org/10.1016/j.ejpe.2016.11.002

Sharifi, M., Kelkar, M., Bahar, A., & Slettebo, T. (2014). Dynamic ranking of multiple realizations by use of the fast-marching method. *SPE Journal*, *19*, 1069–1082. https://doi.org/10.2118/169900-PA

Soam, P. (2019). *Important Points You Should Learn From Oil & Gas Analytics Experts*. Medium. https://medium.com/dataseries/important-points-you-should-learn-from-oil-gas-analytics-experts-7efa48f4c570. Accessed 21 Jan 2021.

Sun, L., & Yan, T. (2019). Technology and application of automated production control robot for petroleum drill pipe. *Academic Journal of Manufacturing Engineering*, *17*, 197–204.

Sun, S., Porth, J., & Panicker, G., et al. (2016). *Methods and Systems for Applying Machine Learning to Automatically Solve Problems*. USA.

Susto, G. A., Schirru, A., Pampuri, S., et al. (2015). Machine learning for predictive maintenance: A multiple classifier approach. *IEEE Transactions on Industrial Informatics*, *11*, 812–820. https://doi.org/10.1109/TII.2014.2349359

Tokarek, T. W., Odame-Ankrah, C. A., Huo, J. A., et al. (2018). Principal component analysis of summertime ground site measurements in the Athabasca oil sands with a focus on analytically unresolved intermediate-volatility organic compounds. *Atmospheric Chemistry and Physics*, *18*, 17819–17841. https://doi.org/10.5194/acp-18-17819-2018

Wang, Z., Di, H., Shafiq, M. A., et al. (2018). Successful leveraging of image processing and machine learning in seismic structural interpretation: A review. *The Leading Edge*, *37*, 451–461. https://doi.org/10.1190/tle37060451.1

White, B., Kreuz, T., & Simons, S. (2018). Midstream. In: Klaus Brun & Rainer Kurz (Eds.) *Compression Machinery for Oil and Gas*. Gulf Professional Publishing Company, Elsevier: Houston, TX, pp. 387–400.

Wrona, T., Pan, I., Gawthorpe, R. L., & Fossen, H. (2018). Seismic facies analysis using machine learning. *Geophysics*, *83*, O83–O95. https://doi.org/10.1190/geo2017-0595.1

Yu, L., Zhao, Y., Tang, L., & Yang, Z. (2019). Online big data-driven oil consumption forecasting with Google trends. *International Journal of Forecasting*, *35*, 213–223. https://doi.org/10.1016/j.ijforecast.2017.11.005

Zainal Abidin, N. W., Ab Rashid, M. F. F., & Nik Mohamed, N. M. Z. (2019). A review of multi-holes drilling path optimization using soft computing approaches. *Archives of Computational Methods in Engineering*, *26*, 107–118. https://doi.org/10.1007/s11831-017-9228-1

Zhang, Y., Bansal, N., Fujita, Y., et al. (2016). From streamlines to fast marching: Rapid simulation and performance assessment of shale-gas reservoirs by use of diffusive time of flight as a spatial coordinate. *SPE Journal*, *21*, 1883–1898. https://doi.org/10.2118/168997-PA

Zhao, Y., Noorbakhsh, A., Koopialipoor, M., et al. (2020). A new methodology for optimization and prediction of rate of penetration during drilling operations. *Engineering with Computers*, *36*, 587–595. https://doi.org/10.1007/s00366-019-00715-2

Chapter 2

AI and ML applications in the upstream sector of the oil and gas industry

2.1 INTRODUCTION

Machine learning (ML) is a type of artificial intelligence (AI) that allows systems to grow and learn without being explicitly programmed. ML is concerned with creating computer programs that can obtain data and information on their own and learn on their own. AI is a machine capable of performing tasks requiring natural intelligence to think like humans or animals (Michalewicz & Michalewicz, 2008). Speech recognition, learning, and problem-solving are some examples of AI. The oil and gas industry is facing several challenges and issues in handling information processing (Mearns & Yule, 2009). In various techniques and processes, a vast amount of data will be generated. For improving the oil and gas industry's performance, the database should be technically and statistically analyzed. The data analysis and interpreting process can be done using AI and ML to solve the gas and oil industry's problems (Holdaway, 2014). AI technology is efficient and has human capabilities like understanding, planning, reasoning, communication, perception, and working effectively at a low cost. This study described how AL and ML learning techniques are used in different upstream industries, their advantages, and disadvantages, particularly in the oil and gas industry, and how ML and AI affect the industry's proper computational and potent decision-making.

The oil and gas business has made fewer investments on AI and ML methods than other industries (Lu et al., 2019). The oil and gas field faces many problems, including rigorous environmental sustainability, equipment maintenance, frequent downtimes, and disconnected locations (Aminzadeh, 2005; Solanki et al., 2021). Although AI has a vast application in different fields, data science and ML are the two major applications of AI in the oil and gas industry (Solanki et al., 2021). ML allows the industries to collect and gather all the information about how oil and gas are discovered, developed, and converted to a data set as a useful insight. Offshore oil and gas sectors can employ AI technology in data science to transform the oil and gas industry's complicated data into a profitable and accessible manner. It creates a new pathway to take advantage of new exploration opportunities beyond the existing infrastructures. Oil and gas industry uses AI to increase the productivity and the efficiency of exploration operations. Entrepreneurs in the upstream oil and gas industry can save up to $50 billion using ML and predictive analytics (Shultz, 2004). Converting the large amount of the data collected to usable and valuable forms using AI and ML helps the companies to improve the return on assets, deliver the returns that investors require, and manage the risks.

DOI: 10.1201/9781003279532-2

2.1.1 Upstream and downstream in the oil and gas industry

The upstream and downstream of oil and gas productions refer to the oil and gas industry location in the supply chain. Typically, the gas and oil sectors are categorized into three categories: upstream (Hassan et al., 2020; Solanki et al., 2021), downstream (Patel et al., 2020), and midstream (Eissa, 2020).

The company is defined as combining the functions of two or three of these groupings (MacKay et al., 2021). Upstream oil and gas production refers to companies that identify, extract, or create raw resources for industrial use. In contrast (Ghaithan et al., 2017), downstream oil and gas production refers to companies that are closer to the customers (Hassani & Silva, 2018). Midstream is a term that encompasses both upstream and downstream services, like transportation and storage.

The upstream sector of the oil and gas business refers to the production and exploration stages. The employees in this part are primarily geologists, service rig operator's scientists, geophysicists, engineering firms, and drilling contractors. These individuals are capable of locating and estimating reserves before any drilling activity begins (Liaw, 2015). China National Offshore Oil Corporation and Schlumberger are two significant companies specializing in upstream services. However, the leading diversified oil and gas companies, such as Exxon-Mobil, account for most of the largest upstream operators (Panico et al., 2017).

Large oil and gas corporations have offices in different places throughout the world. Access to data records may be an essential aspect of their business activities (Tan et al., 2016). Before these data records can be utilized appropriately, they may need to be digitized and evaluated to detect problems such as incomplete or missing information or uncategorized records (Solanki et al., 2021). AI technologies may support oil and gas firms in digitizing records and automating the interpretation of geological data or charts, potentially leading to the discovery of problems such as pipeline corrosion and increased equipment utilization. It is now obvious that society and business are significantly impacted by digital technology. With the passage of time, it became clear that online transformation (Kshirsagar & Shah, 2021) is now considered the "fourth industrial revolution," characterized by the advancements in technologies that blur the distinctions between the physical, digital, and biological realms, like AI robotics and self-driving vehicles. AI technologies are gaining prominence owing to their fast reaction times and great generalization capabilities. Numerous ML techniques utilized in reservoir engineering fall under the category of supervised learning.

2.2 NEED FOR ML-BASED TECHNIQUES IN THE OIL AND GAS UPSTREAM SECTOR

The oil and gas sector has several challenges and difficulties when it comes to data management and processing. To produce a huge amount of data banks, many techniques and processes are employed. A detailed technical evaluation of this information is necessary to improve the performance of the oil and gas industry. So, Sircar et al. (2021) present a comprehensive state-of-the-art assessment of AI and ML for tackling problems in the oil and gas business (Ali et al., 2020). They also explain the many ML and AI approaches used for data processing and analysis in various oil and gas sectors. The advancement and achievements highlight the advantages of AI and ML approaches toward huge data storage capacity and high numerical analysis efficiency.

This paper offers a survey of many scholars' studies on AI and ML applications and constraints for the upstream in oil and gas industrial sectors.

They have indicated that when fossil fuel prices rise, they will have to develop new technologies and enhance operations to optimize production and expand their present capabilities. In addition, other chemicals, such as fertilizers, pharmaceutical drugs, polymers, solvents, and insecticides, need energy sources such as oil and gas. The reservoir sector of engineering uses some of the ML algorithms used in the supervised learning classification. Fuzzy logic (Vilela et al., 2018), supporting vector machines (Benmahamed et al., 2017), artificial neural networks (ANN) (Braimah, 2020), and response surface models (Singh & Kumar, 2020) are some of the advanced machine-learning methods that have been studied. A thorough understanding of blockchain technology (Ajao et al., 2019) will help solve the challenges, opportunities, innovations, and dangers that this industry faces in the oil and gas industry (Wan Ahmad et al., 2016; Wanasinghe et al., 2020).

So that ML does not disturb the physical behavior of the system, it is required to plot or analyze these data using technological intervention and analysis (Tang et al., 2017). As a result, the risk factor and maintenance expenses may be efficiently reduced with this huge embedded device. In addition, the advancement and expansion of developing technologies have become intelligent, making the decision-making process simple. Innovative engineering solutions that are faster, cost-effective, and easier to adopt are becoming increasingly important due to rapid industrialization and technological advancements. One concern is the increased frequency of accidents caused by gas leaks in chemical plants, coal mines, and household appliances.

So, using multimodal AI fusion approaches (Narkhede et al., 2021) offers a novel method for detecting and identifying gaseous pollutants. The bulk of gases and vapors have no odor, color, or taste, making them difficult to detect with our ordinary senses (Babanezhad et al., 2020). Because sensing based on a single sensor might be incorrect, sensor fusion is necessary for comprehensive and robust detection in a variety of real-world applications. They manually gathered 6,400 gas samples using two different sensors: a seven-semiconductor gas sensor array and a thermal camera (1,600 samples per class for four classes). The method of early fusion is centered on multimodal AI. The network design includes a feature extraction technique for each modality, fusing with a merging layer or a dense layer to create a single outcome for gas detection.

They achieved 96% testing accuracy (for the fused model), compared to 82% (Long Short Term Memory (LSTM) was used to analyze data from gas sensors) and 93% (for separate models). The results show that combining modalities surpasses the results of a single sensor, which may also be a group of sensors. Koroteev and Tekic (2021) look at how AI affects the oil and gas business, a big element of the energy sector. They concentrate on the upstream section of the oil and gas industry since it is the most capital-intensive and has the most significant number of uncertainties. They discuss contemporary developments in building AI-based techniques and assess their influence on advancing and de-risking processes in the industry, based on a research of AI application possibilities and an evaluation of existing literature. In the paper, they look into AI techniques and algorithms and the role of data availability. They also look at non-technical barriers to AI's widespread adoption in the oil and gas industry. People, data, and new collaboration are examples.

Thus, if AI is still a new trend in the petroleum industry, there are already applications that have proven to be valuable. They've offered several instances of how AI may assist speed up and reduce the risk of several commercial techniques in the oil and gas sector, including field development, hydrocarbons exploration, or raw hydrocarbon manufacturing.

2.3   OPERATIONS OF AI- AND ML-BASED OIL AND GAS FIELDS

Sircar et al. (2021) have reported the AI and ML techniques used in the oil and gas industry. One of the statistical methods is linear regression, in which the variables evaluated in it should correlate. Global oil production is forecasted using linear and nonlinear regression models. In comparison to other methodologies, the inverse regression model performed better in the oil and gas companies.

The collection and study of data collected before and after drilling into Earth are compulsory for oil and gas industry. Before one spends a hefty amount of money digging into an ineffective well and improving their work efficiently in their regular operations, they want knowledge on solving the production and complex exploration problems. Exploration of hydrocarbons is hazardous. Explorationists must accurately identify subsurface potential to drill and utilize hydrocarbons. Limited 2D seismic data was utilized in the early twenty-first century to locate drilling locations, which relied on underlying mapping. The rate of success was one in seven due to the danger. With time, new data was obtained in each of the leases chosen for study. With advancements in seismic or well data interpretation, processing this massive quantity of data became known as "big data." It took up terabytes of memory space. The ML concept was utilized to examine the enormous amount of data to improve the signal-to-noise ratio. Several robust techniques were employed to analyze the clean data in 2D, 3D, and 4D seismic.

Drilling has several issues, including loss of circulation, stick sleep vibrations, bit wear, borehole instability, and excessive torque. ML can solve these issues (Noshi & Schubert, 2018). Aliouane and Ouadfeul (2014) suggested an ML approach for generating a Poisson's ratio map, which is beneficial for determining rock and drilling direction features. He utilized ML to inspect the quality of huge volumes of drilling data, extract critical data, and forecast downtime. This strategy saved money by reducing the time it took to check vast amounts of drilling data quality. As a result, ML can fundamentally alter the myriad vital decisions made by gas and oil managers and engineers.

The reservoirs in most greenfields are complex in the context of geological and geometrical features. The latter necessitates the construction of high-tech wells with horizontal sections as well as multilateral completions. The good construction is the most expensive operation in field development. So, Koroteev and Tekic (2021) propose the use of all drilling sensor information is essential for maximizing the return on investment. They want to ensure that the wellbore has the best possible contact with the productive part of the formation here. The entire well construction process proceeds at a maximum rate with the least risk of failure and, therefore, with the least amount of downtime.

Drilling in the 21st century is a data-driven process (Maasz, 2020). Sensors come in three varieties. In the first place, there are sensors at the surface that record the drilling process' mechanical parameters in real-time. Another option is to use logging while drilling (Zhong et al., 2020) sensors to measure the creation behind the drilling bit. Sensors that record mechanical information from the bottom hole arrangement are the third type of Measurement While Drilling (MWD) sensor. Using these sensors, it is feasible to manage drilling operations and update the oilfield's geophysics or reservoirs model. There are various techniques for faster, safer, and more precise drilling. AI-assisted drilling support systems that use real-time drilling data are expected to reduce non-productive hours by 20%–40% on average or failures by 90%.

2.4   ACCURATE MODELING AND SMART RIG OPERATIONS

Predictive process models are essential operational tools that allow for more efficient process design and operations. Predictive modeling is getting more popular, and it is precious when we need to use historical data to predict what might occur in the future. Guo (2018) describes a predictive modeling problem with the following characteristics: Several variables (called features) have predictive power and correspond to the variable to be predicted (response). A metric can be used to measure the "accuracy" or "quality" of a prediction. The processes involved in many engineering applications may be too complex to analyze using typical equations based on first-principles models (i.e., physics, chemistry). There may be enough measurable data points to adopt a data-driven method. Instead of memorizing the engineering equations that define a process, we can solely predict the response with the features provided.

According to Cheung (2020), the impact of ML on discovery processes in the gas and oil industry is one of the most noticeable. ML applications in oil and gas allow computers to analyze large volumes of data rapidly and reliably. It includes the ability to discriminate between noise and signal in seismic data. After this data is gathered and analyzed, current software techniques may build accurate geological models. It enables staff to forecast what is under the surface before drilling begins.

It is predicted that if this technology is used across the industry, the number of dry wellheads will be decreased by 10%. The Dutch Central Graben in the North Sea is one current use of ML in the oil and gas industry (Silva et al., 2019). By using ML in the oil and gas industry, engineers could auto-track a Jurassic seismic horizon. Only a few hand seed points were used to accomplish this process. The most recent generations of algorithms produce findings that are more thorough and accurate than any earlier modeling. When asked to analyze tough terrain, these algorithms maintain their accuracy. Faults and stratigraphically problematic areas can be precisely mapped. A laboratory drilling rig has a top drive to set the maximum torque and rotary speed control over a driver. A complete hoisting system will be installed, which will include actuators, brakes, and stepper motors. The top plate is used to organize the top drives and numerous other components that are placed between π load cells. It is connected to the actuator through a coupling, which provides sufficient stability and lifting force. The two pumps are connected through a simple circulation system. The pumps in the circulation system operate at a maximum pressure of 3.1 bar as well as a maximum flow rate of 19 L/min. This rig can conduct vertical good drilling tests in autonomous mode, adaptive advisory system to optimize, associated with surface sensors and high-speed reliable downhole, and has a data management system to process, analyze, visualize and store the data.

In order to build models to differentiate between non-production time and drilling activities like tripping, nine experiments were carried out to gather data on three rig operations (He et al., 2021). Tripping down, tripping up, and rotating on the bottom are also the three procedures in question. The tests include facts on each operation, including circulation, bit rotation, and both operations. As a result, the ML technology, decision tree, supporting vector machine, gradient boosting, and random forest models can recognize various rock forms and rig operations with great accuracy.

2.5   SENSOR-BASED MODELS FOR WELL SPOTTING THE LOCATION

Adegboye et al. (2019) have narrated different pipeline leak detection types, grouped into three categories. Exterior methods are the first category that tracks the external parts of pipelines using specially designed sensing systems. This category includes fiber optic sensors, infrared thermography, ground penetration radar, acoustic emission sensors, and vapor sampling. The second category is detecting leakages in the pipeline through the visualization method. Experienced personnel, helicopters/drones, trained dogs, and smart pigging are included in this category. The third category is the interior method leakage detection using hydrocarbon fluid. This category analyses negative pressure waves, digital signal processing, mass–volume balance, dynamic modeling, and pressure point.

Corrosion has long been a cause of worry in the oil and gas industry, costing the United States billions of dollars each year. The ability to analyze corrosion in real-time, before structural stability is jeopardized, may have a major impact on preventing catastrophic corrosion events. So, Lu et al. (2018) compare conventional and novel corrosion sensors in terms of sensor design, detecting principles, advantages, and limitations. Several corrosion sensor systems were developed for various forms of corrosion, each based on a particular sensing principle. They are classified into two broad categories: direct and indirect corrosion sensors. Direct corrosion sensors directly monitor corrosion processes/rates caused by diverse corrosion sources and corrosive conditions. Indirect corrosion sensors monitor corrosion through corrosion causes like water, low pH, CO_2, leak vibration, wall thickness changes, and strain change.

Corrosion coupons, electrochemical sensors, magnetic flux electrical resistance probes, leakage sensors, ultrasonic testing sensors, pipeline inspection gauges, and electromagnetic sensors are examples of traditional corrosion sensors (Demo & Rajamani, 2020; Sharma et al., 2021). Passive wireless sensors and optical fiber sensors like RFID and SAW are among the developing sensor technologies (Meribout et al., 2021). Compared to electrical-based sensors, optical fiber sensors offer nondestructive monitoring, small size, long reach, flexibility, lightweight, compatibility, inherent immunity to Electromagnetic Interference (EMI) with enhanced safety in the presence of combustible gas/oil, and optical fiber data connection networks. Small size, cost-effectiveness, simplicity of manufacturing, removal of active power, flexibility to a wide range of applications, and compatibility with wireless telemetry are all advantages of passive wireless sensors.

2.6   AI- AND ML-BASED RISK DETECTION SYSTEM AND IMPROVED DRILLING EFFICIENCIES

Rachman (2019) has reported that AI is revolutionary and disturbed by many industries. He believes that AI will have a broader application in every field in the future. Still, most oil and gas industries do not show interest in adapting the AI because the data in the oil and gas industry are considered more secretive and offensive to share. The analyses and interpretation cannot be processed if the data is not shared with AI.

The inconsistencies in the operational integrity of oil and gas industry assets were detected during inspection and maintenance (Jansen Van Rensburg et al., 2019). During the assessment of inspection and maintenance, the patterns will be recognized, and if any variability occurs, they will be noted and omitted. This pattern recognition is the primary work for ML and it can be done effectively and faster than humans. To enable this and identify the anomalies, the selected ML algorithms have to train by labeling the corresponding data as defective or non-defective. Deep learning is one of the programs developed in AI-based artificial neural networks. The oil and gas industry will be operated 24/7.

Therefore, it is necessary to monitor it full time. For instance, the oil and gas industry should monitor the oil leakage in the pipeline. For that, various operating parameters like temperature, pressure, and flow rate should be monitored to identify the anomalies that may cause any failures or leakages. However, monitoring more than three to four parameters is tedious and confusing to humans, so AI is effective. It can be a part of the surveillance system for the oil and gas industry's assets. This paper explains the achievements and developments held by the AI and ML techniques in the different sectors in the upstream oil and gas industry depicted by Sircar et al. (2021). Loss of circulation, stick sleep vibrations, excessive torque, borehole instability, bit wear, etc. are some of the challenges the drilling engineering faces. The Poisson's ratio map is being used to deduce information about the rock's properties and drilling direction that was part of the ML approach. To acquire critical information, verify the reliability of extensive drilling data, and forecast non-productive time, an ML method was used, which aided in lowering labor costs associated with monitoring the quality of extensive drilling data. The Bayesian network may be utilized for managed pressure drilling operations in deep-water drilling. The Bayesian network may be used to forecast failures and do risk assessments in the offshore sector. Automation was utilized to control drilling parameters such as rotary speed, bit weight, and rate of penetration. The Poisson's ratio map is used to deduce information about the rock's properties and drilling direction, which was part of the ML method. To acquire critical information, verify the quality of extensive drilling data, and forecast non-productive time, an ML method was used, which aided in lowering labor costs associated with monitoring the quality of extensive drilling data.

An ML method may gather information such as an alternative bit or rig equipment upgrades, estimated abrasive wear, and anticipated bit wear. As a result, ML can significantly change the numerous crucial decisions by oil and gas administrators and engineers every day.

Real-time drilling risk detection is essential for predicting potential drilling accidents, analyzing their fundamental causes, assessing the amount of risk involved, recommending preventative or control actions, and adjusting control settings as needed to avoid the scenario. The signal changing tendency rate automated extraction method was named after this improved recognition approach by Li et al. (2013). By using a fuzzy (or case-based) reasoning method and a base comparison of live feed with databases reference sets, it is possible to forecast drilling risk in real-time and utilize this information for downhole surveillance of control parameters.

2.7   MACHINE LEARNING-BASED DATA ANALYSIS OF THE WELL LOCATION

Han and Kwon (2021) show how to forecast cumulative gas output. The dependent and independent variables are widely classified into four kinds. They are well information, hydraulic fracturing, completion, and production data, which were evaluated using a deep neural network. Using data-driven deep learning techniques, this study aims to provide a way for building a proxy model to estimate cumulative gas production from hydraulically fractured horizontal drilling in a shale reservoir. Hydraulic fractured and well completion datasets contain both category and numerical variables, as well as production operation data depending on well parameters and data features, and location. The quick and sophisticated degree of the analytic model was developed using complicated analysis techniques. This mathematical approach necessitates the use of a basic spreadsheet to calculate a single-input parameter. The purpose of this article is to use ML modeling to determine the relationship between the input and output data. Using principal component analysis, which extracts significant information from data sets, forecasting performance was excellent when an integral with a cumulative contribution of 85% was used, as well as when a deep neural network model with six variables calculated using variable importance analysis was used.

Cross (2020) classifies oil and gas wells based on geology and engineering characteristics and then takes an average of the group to predict. At computerized speed, the decision-tree-based ML algorithms are working. These algorithms using the oil well data seek the best way to predict similar group production wells. Decision-tree-based models initially start with the trained data in a single group, and then the algorithm will group production wells with similar output and split according to its feature. For understanding and forecasting water production, level ML models are a powerful tool used in the oil and gas industry. These models are a standard Novi product. The advanced level of ML methods developed for oil can also be applied to water. These methods help to solve the problems like complicated interactions of spacing, geology, and completions. It publishes software-driven from the oil and gas production sector. In terms of data processing and management, the oil and gas industry faces several problems and issues. Various approaches are used to produce a large volume of data banks. The different techniques are as follows:

i.   **Artificial Neural Network:** ANN is an ML approach used to address complex issues (Mahmoud & Qurbanov, 2018). ANN is most commonly used in the oil and gas industry to address nonlinear and complicated issues that cannot be handled using a linear connection.
ii. **Linear Regression:** The actual well logging data is interpreted using several linear regression models. The model proved successful in detecting the oil and gas strata by pattern (Peng et al., 2017). Wang (2017) conducted regression research on influencing variables on crude oil's future economy.

Because of their ability to handle large amounts of data and compute quickly, the oil and gas sector is well positioned to profit from ML. If suitable approaches are utilized to incorporate diverse data structures and transform them into usable information that leads to intelligent judgments, the future benefits of data can be realized.

2.8 DIGITAL MODELS FOR THE EXTREME PAPERWORK IN THE FIELD

The study highlights the advantages of cloud data management solutions (Bello et al., 2014). In addition, Bello et al. discuss the most recent market innovations and its plans for the future in terms of cloud data management in general. It also looks at the present obstacles and risks that cloud data management systems face and the business opportunities. Most organizations are presently using virtualization to reduce their computer expenditures. The development of cloud data management systems has resulted from the demand for decreased computing costs. Cloud computing enhances computing by boosting utilization, decreasing maintenance costs and infrastructure, and greater flexibility to meet changing corporate demands.

Hajirahimova (2015) explains the challenges and opportunities of big data analysis in the oil and gas industry. LIS, videos, PDFs, CSV, DLIS, SEGx, Docs XML, etc. are the various data formats. Digital fields, remote operations, reservoir modeling, drilling analyses and predictive plant, and seismic imaging are four practical uses for big data in oil and gas industry. By adopting key success characteristics and avoiding common mistakes, industry profit can get a leg up on the competition. According to the findings, tremendous value can be captured if done correctly. This research paper by Hajirahimova examines the evolution of these methodologies, including their fundamentals, theory, applications, and historical adoption and implementation in the industry. For various areas of application, major problems in the process of implementing these technologies are examined. Compared to traditional methods, the benefits and limitations of data-driven procedures are highlighted and published by Balaji et al. (2018).

2.9 AN EFFECTIVE METHOD FOR PROVIDING DATA TO THE ON-FIELD ENGINEER

By using cutting-edge data-driven techniques, Balaji et al. (2018) demonstrated that predictive modeling, process control, and optimization have all benefited from the use of data-driven models. Due to a lack of well-organized data with just a memory space of about terabytes, the data is usually not utilized properly to analyze and become a valuable insight. Although geology and physics are rarely parts of this data-driven methodology, the industry is not showing interest and still doubting this since it is a data-based solution, not a traditional physics-based solution. This paper assesses the status of the data-driven method with the updates and applications in the oil and gas industry. The best data-driven methods should require a physics-based conventional method, in-depth knowledge of petroleum engineering, traditional statistics methods, AI, data mining, and ML. Data-driven techniques include linear regression, support vector machines, decision trees, fuzzy logic, artificial neural networks, GA, and Bayesian belief networks.

Nguyen et al. (2020) narrated that big data is called voluminous data sets. This paper explains the systematic view of big data analytics in the oil and gas industry. Nowadays, big data technologies have grown rapidly, improving the maximized asset potentials and operational efficiency. The size of data is approximately in either petabyte (=1,024 terabyte) or exabyte (=1,024 petabyte). The data analyst's work is to extract a useful insight, which will predict needful decision-making in various industry sectors. In the oil and gas industry, big data supports monitoring and by comparing new data with old, pipeline systems are maintained. The same techniques are used for maintenance equipment in other segments. In the digitalization of the oil and gas industry, big data analysis is one of the complicated tasks. Installing the most recent generation of Big Data (BD) technology has improved operational efficiency while also increasing asset value. BD development platforms, tools, and network architecture were also examined as part of this study. Increasing computational capacity due to recent developments in data processing, storage, and cloud technologies has made it possible to monitor oil fields and refineries in real-time.

Storing massive quantities of data collected by a shale drilling site was formerly prohibitively expensive (Tahmasebi et al., 2017). Both on-premises and cloud storage costs have dropped significantly. On-premises storage is usually located on a server-class PC that is hooked to the monitoring PC (Syed et al., 2021). One of several common time-series databases, such as OSIsoft Pi, is installed on the server-class PC. Time-series databases, unlike relational databases, are capable of efficiently storing large volumes of real-time data.

Risse (2019) claims that data kept on-premises is frequently required in central locations, like the control center, and could be sent via various methods, including satellite networks and cellular networks. Information from a local computer surveillance system may also be transmitted directly to the cloud, which has several advantages over on-premises data storage. As a result, storage costs per unit are reduced, and storage may grow as needed. Data on the cloud may be accessed from anywhere in the globe using any Internet connection.

2.10 CHALLENGES FACED NOW AND FUTURE SCOPE OF MORE DEVELOPMENT

Nguyen et al. (2020) examined the benefits and disadvantages of big data analytics in the oil and gas sector. The research uses examples from previous BD implementations in Oil and Gas. They explained that technology demands significant financial expenditures and concerted efforts at all business, law, and government levels. Even though the oil and gas industry is used to processing large volumes of data, combining BD analytics with current systems presents several technological and non-technical obstacles. The utilization of current software tools and hardware computer platforms to successfully implement BD technologies is a technical issue. There is no perfect BD employment model that ensures a high-profit improvement while working within time and budget limitations. As a result, future research will look at the possibilities of combining BD with other modern technologies in a unified environment.

Koroteev and Tekic (2021) looked into AI methods and algorithms and the function and data availability in the market. We also go through the non-technical issues preventing the widespread use of AI in the oil and gas sector, such as people, data, or new kinds of association. Companies in the oil and gas industry should rethink their approaches to collaborating with colleges. The truth is that they're not the only ones. Industries (including the oil and gas sector) across the board are being encouraged to make the transition from closed to the innovation process and from collaborations to ecosystem approaches in order to prosper in the AI era (Table 2.1).

Table 2.1 Comparative analysis of the studies

Year	Contributions	Results
2021	• Sircar et al. (2021) explain the unique features of AI and the different methods of ML techniques such as ALM, ANN, fuzzy, supervised learning, logic, PCA, and linear regression. It also displays the importance and the applications of AI and ML in the oil and gas industry.	• ML algorithms like ALM, ANN, and others can dramatically change the important decisions made every day by engineers' administrators in the oil and gas industry and help mature profitable strategies.
2020, 2018	• Guo (2018) used various ML algorithms to forecast the amount of water based on characteristics, including k-nearest neighbors, recurrent neural network with long short-term memory, and gradient boosting machine. • Løken et al. (2020) built models to differentiate between drilling or non-production time activities like tripping; nine tests were undertaken to collect data on three rig operations. Tripping down, tripping up, and rotating on the bottom are also the three procedures in question. • Lu et al. (2018) evaluated traditional corrosion sensors like electrical-based sensors, optical fiber sensors, and new sensor technologies concerning sensor designs, sensing principles, benefits, and limits.	• Light GBM has performed better in terms of predictive power. A new formulation of the issue or more careful tweaking of parameters may have improved the performance of the LSTM algorithm. The ML models can recognize various rock forms and rig operations with great accuracy. • Compared to electrical-based sensors, optical fiber sensors offer non-destructive monitoring, small size, long reach, flexibility, lightweight compatibility, inherent immunity to EMI with optical fiber data communication systems, and enhanced safety when working with combustible gas/oil.
2021	• The Poisson's ratio map identifies rock characteristics information and the drilling direction, which is part of the ML technique (Sircar et al., 2021). The ML method was used to gather important data, examine the quality of huge amounts of drilling data, and estimate the amount of time the well would be inactive. • The goal of this paper is to present a technique for using data-driven deep learning techniques to build a proxy model for forecasting the cumulative gas output using hydraulically fractured horizontal wells in a shale reservoir (Han & Kwon, 2021).	• In the end, the use of ML methods helped save money by reducing the amount of time needed to inspect vast drilling data. The Bayesian network may be utilized in deep-water managed pressure drilling operations. • Employing principal component analysis, which identifies important information from data sets, and forecasting using a deep neural network model using six components computed using variable importance analysis, the prediction performance was shown to be superior.

(Continued)

Table 2.1 (Continued) Comparative analysis of the studies

Year	Contributions	Results
2014, 2018	• Bello et al. (2014) describe obstacles and risks that cloud data management systems face and the business opportunities that exist. He also explains the latest technologies available in the market and their future methods about the general cloud data management technologies. • Balaji et al. (2018) assessed which data-driven method is used to analyze, predict, control, and optimize several processes with the updates and applications in the oil and gas industry.	• Finally, cloud computing is better since it increases use, lowers high installation and maintenance costs, and increases capacity to meet changing business requirements. • It concludes that data-driven techniques will be effective, including linear regression, genetic algorithms, support vector machines, decision trees, artificial neural networks, fuzzy logic, and Bayesian belief networks.
2020	• Nguyen et al. (2020) examine the benefits and disadvantages of big data analytics in the oil and gas sector. The oil and gas industry is used to processing large volumes of data, combining BD analytics with current systems presents several technological and non-technical obstacles.	• There is no perfect BD employment model that ensures a high-profit improvement while working within time and budget limitations. As a result, future research will look at the possibilities of combining BD with other modern technologies in a unified environment.

REFERENCES

Adegboye, M. A., Fung, W.-K., & Karnik, A. (2019). Recent advances in pipeline monitoring and oil leakage detection technologies: Principles and approaches. *Sensors*, *19*(11), 2548. https://doi.org/10.3390/S19112548

Ajao, L. A., Agajo, J., Adedokun, E. A., & Karngong, L. (2019). Crypto hash algorithm-based blockchain technology for managing decentralized ledger database in oil and gas industry. *J*, *2*(3), 300–325. https://doi.org/10.3390/j2030021

Ali, J. A., Kalhury, A. M., Sabir, A. N., Ahmed, R. N., Ali, N. H., & Abdullah, A. D. (2020). A state-of-the-art review of the application of nanotechnology in the oil and gas industry with a focus on drilling engineering. *Journal of Petroleum Science and Engineering*, *191*, 107118. https://doi.org/10.1016/J.PETROL.2020.107118

Aliouane, L., & Ouadfeul, S. A. (2014). Sweet spots discrimination in shale gas reservoirs using seismic and well-logs data. A Case Study from the Worth Basin in the Barnett Shale. *Energy Procedia*, *59*, 22–27. https://doi.org/10.1016/J.EGYPRO.2014.10.344

Aminzadeh, F. (2005). Applications of AI and soft computing for challenging problems in the oil industry. *Journal of Petroleum Science and Engineering*, *47*(1–2), 5–14. https://doi.org/10.1016/J.PETROL.2004.11.011

Babanezhad, M., Taghvaie Nakhjiri, A., Rezakazemi, M., Marjani, A., & Shirazian, S. (2020). Functional input and membership characteristics in the accuracy of machine learning approach for estimation of multiphase flow. *Scientific Reports*, *10*(1), 1–15. https://doi.org/10.1038/s41598-020-74858-4

Balaji, K., Rabiei, M., Suicmez, V., Canbaz, C. H., Agharzeyva, Z., Tek, S., Bulut, U., & Temizel, C. (2018). Status of data-driven methods and their applications in oil and gas industry. *Society of Petroleum Engineers: SPE Europec Featured at 80th EAGE Conference and Exhibition 2018*. https://doi.org/10.2118/190812-MS

Bello, O., Srivastava, D., & Smith, D. (2014). Cloud-based data management in oil and gas fields: Advances, challenges, and opportunities. *Society of Petroleum Engineers: SPE Intelligent Energy International 2014*, 652–660. https://doi.org/10.2118/167882-MS

Benmahamed, Y., Teguar, M., & Boubakeur, A. (2017). Application of SVM and KNN to Duval Pentagon 1 for transformer oil diagnosis. *IEEE Transactions on Dielectrics and Electrical Insulation*, *24*(6), 3443–3451. https://doi.org/10.1109/TDEI.2017.006841

Braimah, M. N. (2020). Application of Artificial Neural Network (ANN) in the optimization of crude oil refinery process: New Port-Harcourt refinery. *Journal of Energy Research and Reviews*, 26–38. https://doi.org/10.9734/JENRR/2020/V5I430154

Cheung, K. C. (2020). *10 Applications of Machine Learning in Oil & Gas–Algorithm-X Lab*. Algorithm-X Lab. https://algorithmxlab.com/blog/10-applications-machine-learning-oil-gas-industry/

Cross, T. (2020). *Oil Well Data Analysis with Machine Learning for Wells in Production*. Novilabs. https://novilabs.com/oil-and-gas-well-data-analysis-for-wells-in-production/

Demo, J., & Rajamani, R. (2020). Corrosion Sensing. *Structural Integrity*, *13*, 83–104. https://doi.org/10.1007/978-3-030-32831-3_3

Eissa, H. (2020). Unleashing industry 4.0 opportunities: Big data analytics in the midstream oil & gas sector. *International Petroleum Technology Conference 2020, IPTC 2020*. https://doi.org/10.2523/IPTC-19802-ABSTRACT

Ghaithan, A. M., Attia, A., & Duffuaa, S. O. (2017). Multi-objective optimization model for a downstream oil and gas supply chain. *Applied Mathematical Modelling*, *52*, 689–708. https://doi.org/10.1016/J.APM.2017.08.007

Guo, J. (2018). *Predictive Modeling using Machine Learning. Process Ecology*. https://processecology.com/articles/predictive-modeling-using-machine-learning-in-the-upstream-oil-gas-sector

Hajirahimova, M. S. (2015). *Review of Statistical Analysis Methods of Large-Scale Data*. https://doi.org/10.1109/ICAICT.2015.7338519

Han, D., & Kwon, S. (2021). Application of machine learning method of data-driven deep learning model to predict well production rate in the shale gas reservoirs. *Energies, 14*(12), 3629. https://doi.org/10.3390/EN14123629

Hassan, A., Mahmoud, M., Bageri, B. S., Aljawad, M. S., Kamal, M. S., Barri, A. A., & Hussein, I. A. (2020). Applications of chelating agents in the upstream oil and gas industry: A review. *Energy & Fuels, 34*(12), 15593–15613. https://doi.org/10.1021/ACS.ENERGYFUELS.0C03279

Hassani, H., & Silva, E. S. (2018). Big Data: A big opportunity for the petroleum and petrochemical industry. *OPEC Energy Review, 42*(1), 74–89. https://doi.org/10.1111/opec.12118

He, J., Chen, W., Chen, L., & Shi, C. (2021). Research and development of drilling operation data benchmarking management system. *Journal of Physics: Conference Series, 1894*(1), 012102. https://doi.org/10.1088/1742-6596/1894/1/012102

Holdaway, K. R. (2014). *Harness Oil and Gas Big Data with Analytics: Optimize Exploration and Production with Data Driven Models*. https://books.google.com/books/about/Harness_Oil_and_Gas_Big_Data_with_Analyt.html?id=hwaIAwAAQBAJ

Jansen Van Rensburg, N., Kamin, L., & Davis, S. (2019). Using machine learning-based predictive models to enable preventative maintenance and prevent ESP downtime. *Society of Petroleum Engineers: Abu Dhabi International Petroleum Exhibition and Conference 2019, ADIP 2019*. https://doi.org/10.2118/197146-MS

Koroteev, D., & Tekic, Z. (2021). Artificial intelligence in oil and gas upstream: Trends, challenges, and scenarios for the future. *Energy and AI, 3*, 100041. https://doi.org/10.1016/J.EGYAI.2020.100041

Li, X., Chan, C. W., & Nguyen, H. H. (2013). Application of the neural decision tree approach for prediction of petroleum production. *Journal of Petroleum Science and Engineering, 104*, 11–16. https://doi.org/10.1016/j.petrol.2013.03.018

Liaw, A. (2015). SEG Deepwater exploration workshop, Shenzhen, China. *The Leading Edge, 34*(12), 1520–1521. https://doi.org/10.1190/TLE34121520.1

Løken, E. A., Løkkevik, J., & Sui, D. (2020). Data-driven approaches tests on a laboratory drilling system. *Journal of Petroleum Exploration and Production Technology, 10*(7), 3043–3055. https://doi.org/10.1007/S13202-020-00870-Z

Lu, H., Guo, L., Azimi, M., & Huang, K. (2019). Oil and gas 4.0 era: A systematic review and outlook. *Computers in Industry, 111*, 68–90. https://doi.org/10.1016/J.COMPIND.2019.06.007

Lu, P., Morris, M., Brazell, S., Comiskey, C., & Xiao, Y. (2018). Using generative adversarial networks to improve deep-learning fault interpretation networks. *Leading Edge, 37*(8), 578–583. https://doi.org/10.1190/tle37080578.1

Maasz, G. J. (2020). *Increasing Overall Equipment Effectiveness of Drilling Machines by Means of Data Driven Dashboards*. North-West University, South Africa. https://repository.nwu.ac.za/handle/10394/36230

MacKay, K., Lavoie, M., Bourlon, E., Atherton, E., O'Connell, E., Baillie, J., Fougère, C., & Risk, D. (2021). Methane emissions from upstream oil and gas production in Canada are underestimated. *Scientific Reports, 11*(1), 1–8. https://doi.org/10.1038/s41598-021-87610-3

Mahmoud, M. M. A. S., & Qurbanov, Z. (2018). Review of fuzzy and ANN fault location methods for distribution power system in oil and gas sectors. *IFAC-PapersOnLine, 51*(30), 263–267. https://doi.org/10.1016/J.IFACOL.2018.11.298

Mearns, K., & Yule, S. (2009). The role of national culture in determining safety performance: Challenges for the global oil and gas industry. *Safety Science, 47*(6), 777–785. https://doi.org/10.1016/J.SSCI.2008.01.009

Meribout, M., Mekid, S., Kharoua, N., & Khezzar, L. (2021). Online monitoring of structural materials integrity in process industry for I4.0: A focus on material loss through erosion and corrosion sensing. *Measurement, 176*, 109110. https://doi.org/10.1016/J.MEASUREMENT.2021.109110

Michalewicz, Z., & Michalewicz, M. (2008). Machine intelligence, adaptive business intelligence, and natural intelligence. *IEEE Computational Intelligence Magazine, 3*(1), 54–63. https://doi.org/10.1109/MCI.2007.913389

Narkhede, P., Walambe, R., Mandaokar, S., Chandel, P., Kotecha, K., & Ghinea, G. (2021). Gas detection and identification using multimodal artificial intelligence based sensor fusion. *Applied System Innovation, 4*(1), 3. https://doi.org/10.3390/ASI4010003

Nguyen, T., Gosine, R. G., & Warrian, P. (2020). A systematic review of big data analytics for oil and gas industry 4.0. *IEEE Access, 8*, 61183–61201. https://doi.org/10.1109/ACCESS.2020.2979678

Noshi, C. I., & Schubert, J. J. (2018). The role of machine learning in drilling operations; a review. *SPE Eastern Regional Meeting.* https://doi.org/10.2118/191823-18erm-ms

Panico, M., Tang, H., Fairchild, D. P., & Cheng, W. (2017). ExxonMobil SENT test method and application to strain-based design. *International Journal of Pressure Vessels and Piping, 156*, 17–22. https://doi.org/10.1016/J.IJPVP.2017.06.010

Patel, H., Prajapati, D., Mahida, D., & Shah, M. (2020). Transforming petroleum downstream sector through big data: A holistic review. *Journal of Petroleum Exploration and Production Technology, 10*(6), 2601–2611. https://doi.org/10.1007/s13202-020-00889-2

Peng, H., Lima, A. R., Teakles, A., Jin, J., Cannon, A. J., & Hsieh, W. W. (2017). Evaluating hourly air quality forecasting in Canada with nonlinear updatable machine learning methods. *Air Quality, Atmosphere and Health, 10*(2), 195–211. https://doi.org/10.1007/s11869-016-0414-3

Rachman, A. (2019). *How to Apply Artificial Intelligence in the Oil and Gas Industry.* Medium. https://medium.com/swlh/how-to-apply-artificial-intelligence-in-the-oil-and-gas-industry-4fb52e1dd50e

Risse, M. (2019). *Refineries need Machine Learning to Improve Operations.* Oil & Gas Engineering. https://www.oilandgaseng.com/articles/refineries-need-machine-learning-to-improve-operations/

Sharma, V. B., Singh, K., Gupta, R., Joshi, A., Dubey, R., Gupta, V., Bharadwaj, S., Zafar, M. I., Bajpai, S., Khan, M. A., Srivastava, A., Pathak, D., & Biswas, S. (2021). Review of structural health monitoring techniques in pipeline and wind turbine industries. *Applied System Innovation, 4*(3), 59. https://doi.org/10.3390/ASI4030059

Shultz, J. (2004). *Follow the Money: A Guide to Monitoring Budgets and Oil and Gas Revenues.*

Silva, R. M., Baroni, L., Ferreira, R. S., Civitarese, D., Szwarcman, D., & Brazil, E. V. (2019). *Netherlands Dataset: A New Public Dataset for Machine Learning in Seismic Interpretation.* https://arxiv.org/abs/1904.00770v1

Singh, B., & Kumar, P. (2020). Pre-treatment of petroleum refinery wastewater by coagulation and flocculation using mixed coagulant: Optimization of process parameters using response surface methodology (RSM). *Journal of Water Process Engineering, 36*, 101317. https://doi.org/10.1016/J.JWPE.2020.101317

Sircar, A., Yadav, K., Rayavarapu, K., Bist, N., & Oza, H. (2021). Application of machine learning and artificial intelligence in oil and gas industry. *Petroleum Research.* https://doi.org/10.1016/J.PTLRS.2021.05.009

Solanki, P., Baldaniya, D., Jogani, D., Chaudhary, B., Shah, M., & Kshirsagar, A. (2021). Artificial intelligence: New age of transformation in petroleum upstream. *Petroleum Research.* https://doi.org/10.1016/J.PTLRS.2021.07.002

Syed, F. I., Alnaqbi, S., Muther, T., Dahaghi, A. K., & Negahban, S. (2021). Smart shale gas production performance analysis using machine learning applications. *Petroleum Research.* https://doi.org/10.1016/j.ptlrs.2021.06.003

Tahmasebi, P., Javadpour, F., & Sahimi, M. (2017). Data mining and machine learning for identifying sweet spots in shale reservoirs. *Expert Systems with Applications, 88*, 435–447. https://doi.org/10.1016/j.eswa.2017.07.015

Tan, K. H., Ortiz-Gallardo, V. G., & Perrons, R. K. (2016). Using Big Data to manage safety-related risk in the upstream oil & gas industry: A research agenda. *Energy Exploration & Exploitation, 34*(2), 282–289. https://doi.org/10.1177/0144598716630165

Tang, Y., Jing, J., Zhang, Z., & Yang, Y. (2017). A quantitative risk analysis method for the high hazard mechanical system in petroleum and petrochemical industry. *Energies, 11*(1), 14. https://doi.org/10.3390/EN11010014

Vilela, M., Oluyemi, G., & Petrovski, A. (2018). Fuzzy data analysis methodology for the assessment of value of information in the oil and gas industry. *IEEE International Conference on Fuzzy Systems.* https://doi.org/10.1109/FUZZ-IEEE.2018.8491628

Wan Ahmad, W. N. K., Rezaei, J., Tavasszy, L. A., & de Brito, M. P. (2016). Commitment to and preparedness for sustainable supply chain management in the oil and gas industry. *Journal of Environmental Management, 180*, 202–213. https://doi.org/10.1016/J.JENVMAN.2016.04.056

Wanasinghe, T. R., Wroblewski, L., Petersen, B. K., Gosine, R. G., James, L. A., De Silva, O., Mann, G. K. I., & Warrian, P. J. (2020). Digital twin for the oil and gas industry: Overview, research trends, opportunities, and challenges. *IEEE Access, 8*, 104175–104197. https://doi.org/10.1109/ACCESS.2020.2998723

Wang, K. (2017). *Regression Analysis of Influencing Factors on the Future Price of Crude Oil.* https://doi.org/10.24104/rmhe/2017.02.01015

Zhong, R., Johnson, R. L., & Chen, Z. (2020). Using machine learning methods to identify coal pay zones from drilling and logging-while-drilling (LWD) data. *SPE Journal, 25*(3), 1241–1258. https://doi.org/10.2118/198288-PA

Chapter 3

One step further in upstream sector

3.1 INTRODUCTION

The oil and gas (O&G) sector is a significant polluter of the environment (Ali et al., 2020; Kshirsagar, 2018). Manufacturing of O&G, as well as energy consumption, is continuously growing. The expansion and carbon-free economy, increasing effectiveness of regenerative and green energy are the most widely debated topics in the evolution of global energy today. However, in the energy sector, oil business trends are not given adequate consideration, like the sacredness of technical and digital development (Kshirsagar & Shah, 2021). Since 2000, digitization processes have been infiltrating the oil sector as the fourth industrial revolution. Nowadays, the technical solutions in oil production and exploration allow for considerable cost reductions in developing reserves and increased oil supply volume, raising the oil industry's competition to a new high (Maksimov et al., 2017). The upstream O&G companies can quickly separate and analyze a massive quantity of data by integrating drilling rig sensor data and operational data into true windows and utilizing computing tools to mine the data for reservoir development appraisal (Mohammadpoor & Torabi, 2020).

Support vector machines (SVMs) (Orrù et al., 2020) and multiple linear regressions (Balaji et al., 2018) examine the data and gradually apply it to anticipate uncertainty. They can generate crucial bits of information that aid in execution while also predicting difficulties, allowing them to improve drilling and production results. Generally, extensive data analysis provides in the long-term increment of the O&G industry. O&G (petroleum) has remained a significant source of energy for many years. O&G, as well as their products, are consumed by all countries. Both producers and consumers are interested in O&G pricing and derivatives. The price of O&G influences the level of costs in all sectors of production. Because many countries' economies are built on O&G production and trading in O&G products, predicting O&G prices is essential (Rahman et al., 2020). It is important to note that O&G prices directly impact some sectors of the economy. O&G prices affect political and economic processes that influence the stock value of O&G businesses, the inflation rate in O&G importing countries, and economic growth. The progressive depletion of specific big fields found in the year 1960, which, since the 1990s, and therefore the worldwide reduction in the supply of materials, fueled the development of oil production technology. The flow of world reserves grew by 60% from the 1980s to the 1990s, but only by 4% from the 1990s to the 2000s. Russia's O&G output has grown by 5.5% and 13% during the last five years, respectively. Emissions of hazardous substances are a direct result of energy generation.

DOI: 10.1201/9781003279532-3

Technological advancements, automation of manufacturing processes, and modern monitoring devices enable timely collection and neutralization of air pollutants, reducing negative environmental consequences.

Knowledge is a collection of information, data, experience, and adept opinion that helps assess and integrate new information and experience. Sharing the information is crucial to a company's profit, which leads to implementing the information faster in every sector of the company, leading to a profit and improving the industry's performance and competitiveness.

Data and information management is one of the most difficult tasks in the O&G business. Engineers are frequently overloaded by enormous amounts of data, leading them to ignore crucial information that can sometimes help them better comprehend the reservoir. As a result, Petroleum Data Analysis has become increasingly popular. In 2014–2019, the average annual volume of oil consumption was over 4.2 billion tons, which increased 54% from 1974 to 1979. As a result, the average annual rise in O&G consumption since the oil shock was 1%. At the same time, following the 1973–1983 economic crisis, oil consumption increased consistently until the 2008 crisis. Significant and unexpected variations in O&G prices, on the other extreme, are widely believed to have a negative influence on the well-being of both O&G importing countries and O&G producing countries. In 2018, shale gas accounted for over half of all-natural gas production in the United States. Total primary energy supply analysts predict oil at 32%, coal at 29%, natural gas at 23%, biofuel and waste at 10%, nuclear at 5%, hydro at 2%, and others at 0.15%. The third-largest energy source in the world is natural gas. As a result, more wells are being drilled, and new natural gas reservoirs are being searched to supply this need.

3.2   DIGITALIZATION AND AUTOMATION IN EXPLORING SECTOR

The aim of the study by Shinkevich et al. (2020) is to provide a technique for analyzing the Russian economy's long-term growth in the O&G industry while considering digitalization. As part of the methodological basis, the data systematisation technique was used to follow the kinetics of variations in ecological and financial indicators in the O&G sector (Orazalin et al., 2019), as well as the energy sector overall; modelling methods (including principal component methods and correlation–regression), which determine the computational association between the O&G sector's digital transformation and measures of sustainable development; predicting approach that provides scenario for the alterations in the O&G sector's environmental friendliness. As a consequence, a linear regression approach represents the dependency of mining companies' problematic substance emissions on the financial ratio of leverage and defines the need to raise the proportion in order to minimize toxic outcomes; alternate solution prediction for the rate of toxic materials by the Russian economy's mining industry is established. An integrated sustainable growth index is developed using the factor analysis method to analyze the long-term growth of the Russian economy's O&G factory from the perspective of digitization. The elements revealed by principal component analysis allow for an assessment of the impact of digitalization on the long-term growth of Russia's O&G business. The correlation coefficients between each characteristic and the selected factor were also computed. Relying upon the simulation outcome, the interrelations of two main components – the O&G industry's ongoing expansion and its technology trends – are evident. The findings of this study may be included in critical plans and programs for the growth of the O&G industry and the technological finance of the Russian Federation's industrial system.

The purpose of this article is to examine challenges and determine prospects for the digitization of Russia's oil sector (Kulyasov et al., 2020). The research looked at the state of digitalization in the oil sector, the methodologies (such as big data analysis, Internet of Things, and data visualization) employed, and the future orientations of Business 4.0 in the oil industry. The major issues with the growth of the Russian Federation's oil industry's digitization have been highlighted. The absence of clear preferences for technological advancement of the oil industry in Russia, a severe scarcity of funding for early stages of R&D, a lack of development of venture economic investment in hands-on tasks, hindrances, in firms, and other issues are among the major challenges in the advancement of the oil industry digital revolution in Russia (Kontorovich et al., 2017). To balance out the negative aspects, a systematic approach like R&D is suggested, which comprises three main areas for Russia's active technology of O&G production. As a consequence of the study, a methodical strategy was presented that comprises three goals for the growth of the oil industry's digitization in Russia, whose execution would encourage and improve sector R&D, institutional ecology, and venture economic investment. The researchers completed the research tasks using analytical, statistical, conceptual, financial analysis, and system techniques.

Artificial intelligence (AI) approaches are discussed by Kshirsagar and Shah (2021) and Solanki et al. (2021) for effective planning, determining reservoir rock characteristics, mining optimization, and manufacturing facilities. Reservoir engineers can design a solid reservoir development strategy and efficiently control hydrocarbon extraction with accurate knowledge of porosity and permeability. Wireline log seismic and data characteristics are utilized in AI-based algorithms to estimate porosity (NianTian et al., 2018). The modeling of AI approaches to estimate homogeneous liquid permeability has been explored by considering the various forms of well log information. Compared to the actual measurement, the estimation of permeability of the mercury injection capillary pressure information of capillary pressure creates a significant inaccuracy. An artificial neural network (ANN) was constructed to tackle this problem for the real permeability assessment is also described in this article. The drilling fluid is designed and accurately analyzed using AI techniques. It helps with drilling process selection and drilling-related issues. If a lake cannot generate organically, AI may choose the correct artificial method and build a system for production. NAVA and SADA base of knowledge systems are addressed about Enhanced Oil Recovery (EOR) techniques. AI can assist in improving output, increasing efficiency, and increasing hydrocarbon recovery.

3.3 MAPPING AND ANALYZING OF THE FIELD DIGITALLY

Crompton (2015) in his paper discussed that in 2002, Cambridge Energy Research Associates published a report called "Digital Oil Field of the Future." Over the previous ten years, a lot has transpired, some of it amazing, some of it not. This appears to be an appropriate time to review where the industry stands and where it is heading with its Digitized Oil Field/Integrated Process expenditures. Evaluation of Digitized

Oil Field/Integrated Process initiatives is done against the context of Gartner's well-known Technology Innovation Hype Curve (Jabar, 2021). There are numerous huge opportunities in each O&G engineering and geoscience discipline, which includes drilling (true borehole status monitoring and forecasting and prevention of unexpected incidents, particularly well management); completions (evaluation of water fracturing operations, such as different stages of fracturing and microseismic); and preservation and operation (real-time reservoir management). The task at hand is to determine how to maximize the return on a digital petroleum industry investment. However, there are the following obstacles to Digitized Oil Field/Integrated Process adoption: "integration challenge," "innovation challenge," "complexity challenge," "data foundation challenge," and "knowledge exchange challenge," among others. Several aspects to consider are as follows: (1) The pattern of system automated processes, real-world systems, and reservoir and earth modeling implies that the electronic oil sector has become an actuality. (2) Young employees and global geologists are joining the tasks with increased technological skills from lifetime skills with customer IT and some coding graduate studies. (3) Petroleum engineering and earth science are supplying more "intellectual property" through software. (4) Substantial gaps continue to exist.

The magnitude, difficulty, technologies employed, and benefits of adopting the NK-KwIDF are discussed in the paper by Al-Subaiei et al. (2019). North Kuwait (NK) contains 5 fields, over 1,000 operational wells, and 7 gathering centers. To link these 1,200 boreholes to gathering centers, a sophisticated network of pipelines, trunk lines, and manifolds is employed. Each item of NK crop productivity must be handled by maximizing appropriate wells and decreasing downtime in each area, resulting in a massive ground system structure. The extended surface networking model considers all of the field's features, for example, manifolds, pipelines, wells, and details about gathering centers. A well model is created for each well in the NK assets, taking into account all PVT data and surface coordinates, and then coupled to a subsurface network structure with all pipe information. After creating the detailed surface model, various integrated processes were created to run the model and evaluate the results quickly (Hassan et al., 2019). Electrical Submersible Pumps optimizing and Electrical Submersible Pumps analysis processes are capable of identifying oil gain chances and diagnosing Electrical Submersible Pumps functionality. The identified chances are entered into a ticketing system, which tracks the opportunity's life cycle from recognition through execution in oil profits. The NK-KwIDF project produced excellent results in terms of oil yield from the well-level optimization and different network optimization. It developed a strong reputation in the oil business, attracting numerous NOCs and IOCs to come and learn from the flagship's development and deployment of digital field technologies.

3.4   SPOTTING DRILLING AND PIPELINE LOCATION PRECISELY USING ML-BASED APPLICATIONS (TOTAL OIL AND GOOGLE CLOUD)

In the drilling process, stuck pipe events are a typical occurrence that might result in increased waste of time (Al Dushaishi et al., 2021). To avoid or lessen the severity of these events, standard suggested practices are employed. Various non-physical approaches, like

ANNs, have been used to anticipate these occurrences based on observed data. In this study, classification trees were created using recursive partition analysis. The information was gathered from 385 wells bored in various locations throughout southern Iraq. A sum of 1,015 data points was obtained, classified into three sets of data: testing, training, and validation. The primary purpose of the project by Rubani (2017) is to develop a model with simple factual criteria that can estimate blocked pipe events and provide a way to open the pipe. The proposed technique predicted stopped pipe occurrences with 90% reliability using basic and limited input data. The prediction accuracy for releasing a blocked pipe in the stuck pipe treatment model was 84%. Based on logical criteria and component amounts, a recursive partition was able to recognize blocked pipe occurrences. The recommended models for predicting blocked pipe occurrences and remedies give a logical criterion based on fundamental factors that may be implemented in the field.

One of the causes of inadequate time (NPT), which can raise well expenses, is stuck pipes (Khan et al., 2020). This study looks at the possibility of utilizing machine learning (ML) to estimate the probability of stuck pipelines while drilling rigs in oilfields. The estimated model is designed to forecast the possibility of blocked pipes as then appropriate borehole operation employees are alerted and may implement a mitigation strategy to avoid them. SVM (Agwu et al., 2018) and ANN (Shadizadeh et al., 2010) were the two ML techniques investigated in this study. For the well drilling process, a sum of 268 sets of data was successfully gathered using the extraction of data. The parameters that caused the blocked pipes during the bore holing procedure are also included in the data. The characteristics of the operating circumstances, drilling fluid, the status of the borehole, and the BHA specification are all included in these drilling parameters. The SVM and ANN ML models were built using the R programming language. The ML models' prediction performance was assessed in terms of specificity, accuracy, and sensitivity. These two ML methods were subjected to a sensitivity analysis. Aside from that, SVM hyper-parameters, namely the degree (D), sigma (S), and regularization factor (C) employed in the sensitivity analysis. The best ANN model achieved 91.89% sensitivity, 88.89% accuracy, and 86.36% specificity. In comparison, the best SVM model achieved an 86.49% sensitivity, 83.95% accuracy, and 81.82% specificity, according to the sensitivity analysis. Compared to the good SVM model, the ANN model is the best ML model in this paper since its sensitivity, specificity, and accuracy are constantly larger. In summary, based on the promising prediction accuracy exhibited in the findings of this study (Khan et al., 2020), recommended that utilizing ML to forecast blocked pipes is certainly feasible.

3.5 DIGITALLY MONITORED PRODUCTION SITES

Priyanka et al. (2021) intend to present a digital twin framework based on prognostics algorithms and ML and compute and estimate the probability proportion of an oil pipe network. Prognostics is concerned with detecting a breakdown precursor by calculating the risk condition related to pressure data to determine the remaining useful life. The irregularity of the pressure characteristic is used in predictive computation to estimate risk likelihood, and the abnormal pressure decrease and rise are separated using Canopy Clustering and Dirichlet Process Clustering. Features are obtained using the multi-oil substation information management system and various learning methodologies. The optimal feature likelihood rates are assessed using a kernel-based SVM algorithm to give an on-time control system on a whole oil pipe network via effective wireless communication systems between the server and oil substations (Priyanka et al., 2021). Consequently, by merging a complete transmission system with expanded wireless data networks in remote places, the proposed method builds an online Intelligent Integrated Automated Control Method to anticipate the proportion of danger in the oil sector.

The purpose of the article is to describe how to utilize big data technology to maximize operations and demonstrate how big data may be used to obtain crucial operational intelligence in the O&G business and aid in decision-making in various upstream activities (Baaziz & Quoniam, 2013). Meta Group (now part of Gartner) came up with the original description of big data by identifying the three "3V" characteristics: volume, velocity, and variety (Johnson et al., 2017). IBM has established a fourth category (V: veracity) based on data quality. On the other hand, Oracle added a fourth "V," that is value, to emphasize the extra value of big data. By merging the different definitions big data (Baaziz & Quoniam, 2013), it gets a more comprehensive definition termed as the "5Vs" namely veracity, value, variety, velocity, and volume (Prajapati & Patel, 2021). Major O&G firms have already started initiatives to implement big data technology that will allow them to monitor new economic opportunities, save expenses, and restructure operations. O&G businesses generate actual examples of big data applications by realizing the value of underutilized data resources in enabling fact-based decision-making. Then they will be able to produce increased company value based on creativity, which will lead to a long-term competitive edge. Organizations in the gas and oil industry must first do a gap assessment to establish the most critical technology and data-management expertise requirements. This enables concentrated investment in established and proven technologies and those that will deal with exponentially increasing data volumes. O&G businesses must devise innovative ways to manipulate this information and utilize it to assist specialists in their business processes and administrators in decision-making.

3.6   PLANNING AND COMMISSIONING THE ONSHORE AND OFFSHORE PRODUCTION SITE BASED ON THE ML MODELS

An offshore sustainable energy farm's lifetime expenses include a significant portion of capital and operating costs (Rinaldi et al., 2021). A significant variety of approaches and procedures have been created to assist strategy development and resource managerial decision-making. Because of its advantages in defect detection (Fioravanti et al., 2019), performance assessment, and development, process monitoring equipment is widely employed, particularly in offshore wind turbines. Integrating technological developments, the offshore service business is moving toward automation and digitalization. With an emphasis on the offshore wind sector, this article (Rinaldi et al., 2021) analyses the advantages and limitations of existing literature and innovative approaches in supporting and repairing management and quality control of offshore renewable power fields. State of the art in commercial condition-based management and degradation models and damage detection and prognosis approaches are discussed. Future robots, AI, and data processing possibilities are examined. The consequences of initial project incorporation and the implementation problems of these techniques and Industry 4.0 options in the offshore renewables sector are examined. Different techniques must be linked into a unique framework that includes everything from data gathering through multiple sensors to service scheduling via optimization approaches. Most significantly, this approach should be updated regularly utilizing actual data gathering, analysis, and forecasting to modify repair choices proactively in light of current restrictions and situations. Industry 4.0 (Senum, 2021) principles enable this real-time data stream since the virtual and physical sectors need to be closely integrated and harmoniously operate.

Despite the reality that each facility has its requirements, the O&G sector plays a critical role in nation-building (Sabri et al., 2015). Given the large sums of money required to kick-start an O&G project, reasonable efforts appear to be critical in assuring the project's success. One of the greatest answers to this problem appears to be a suitable task management strategy. It is clear that having a standard project maintenance strategy and advice, such as the Project Planning Book of Knowledge and Stage-Gate Project Management Process, is important for sustainable project execution. The application can also be adapted to each country's requirements and business, with certain authorization and project phase's variances between downstream and upstream O&G projects. (Sabri et al., 2015) This article was beneficial in fulfilling its purpose of analyzing project administration in the O&G sector; nevertheless, more research into the specific necessary success aspects that enable the successful execution of O&G projects around the world is likely.

3.7   DIGITALLY GOVERNED PRODUCTION AND STANDARDIZED DATA COLLECTING

According to Saputelli (2017), many companies have reaped significant gains from data-driven analytics in recent decades. The O&G industry is still in its infancy. The benefits include better decision-making quality, better planning and forecasting, cost reductions, and increased operating effectiveness throughout other industries. However, in our field, there are several barriers to broad acceptance. Aside from the old data-management issues, there are significant gaps in knowledge of basic concepts of how and when to utilize various data analytics technologies (Abdalla, 2018). The effective utilization of sources of data to generate insights and enhance decision-making demonstrates data analytics benefits. In recent years, the number of applications for improving data quality after or during acquisition by removing background noise and anomalies as soon as possible, effectively integrating innovative and effective data into science designs, improving schedule verifications for precautionary tasks, and ensuring the availability of well, surface, and drilling system equipment has grown exponentially. Text mining and natural language processing make it possible to rapidly and efficiently extract relevant details from word reports. These methods make operational transaction data (e.g., suggestions, failure/success) captured in the unstructured language available to a previously underused yet useful source. Natural language processing has been effectively utilized in the drilling industry to characterize and forecast time that is not useful and time that is not apparent from a huge quantity of unstructured data gathered from drilling rigs documents (Lun et al., 2021). Production optimization and reservoir management will also make significant contributions.

Organizing information and data is one of the numerous difficulties we encounter daily in the petroleum sector (Saputelli, 2016). Saputelli (2016) conveyed that sometimes influenced by the large amount and variety of formats available, and in several circumstances, then distracted from the information that needs to comprehend a process, forecast the coming years, or make sound judgments. Data analytics, which supports proper engineering and managerial judgment and modeling actual fuel situations, is the solution to these issues. To minimize low-oil-price circumstances, data analytics for strategic planning is continuously being created. Data evidence and patterns have been used more than the first principle modeling techniques in forecasting and prescribing. The exponential rise of data and quantities has been aided by advances in computer power, sensor accessibility, and engineering concepts (Priyanka et al., 2021). To handle such additional difficulties, data-driven approaches also have varied and improved. The administration of enormous and complicated data quantities is now called "big data." The people who manage and extract value from data are referred to as "data scientists." The following groups of data analytics tools may be used: pattern recognition (classification, principal components, regression, ensemble averaging, and Markov models), AI (planning, perception, and creativity), ML (inductive logic programming, clustering, and rule learning), statistics (regression, factor analysis, and time series), business intelligence (multi-dimensional visualization and key-performance-indicator dashboards), and managing the huge number of datasets, cloud computing services, parallel computing, profiling, and data cleansing.

3.8   SCADA-BASED NETWORK FOR EFFECTIVE COMMUNICATION IN THE FIELD

To prevent significant environmental contamination and financial damage caused by leakage, a system for pipeline leakage estimation and tracking must be created to discover leaks and pinpoint leak sites (Wang et al., 2004). Over 98% of pipe leaking in China is paroxysmal; the negative pressure wave propagation technique is an excellent way to detect the leak location. Although there is a SCADA system to supervise the operation of long-distance petroleum pipelines, the functionality of leak estimation and tracking is not implemented in the current SCADA system in China. The DDE technique extracted pipe operating parameters from the SCADA system, such as valve position, pressure, temperature, flow rate, and pump current. This makes use of the SCADA system's extensive data collecting capabilities to offer data for leak detection and tracking (Wanasinghe et al., 2020). Instead of using a system model, the wavelet packet analyzing fault diagnostic approach may immediately detect defects by changing parameters such as the power of the frequency component. This Wanasinghe extracts the feature data of leakage pressure signals using a wavelet packet analysis-based properties removal approach. The eigenvector indexes can be combined with the values acquired from the SCADA system to eliminate false alerts. In this study, wavelet analysis was utilized to precisely pinpoint leak locations (Wang et al., 2004).

The article by Vijay and Unni (2012) discusses the recent development of SCADA and its surroundings and the need for enhanced security in these networks. It also discusses the new study topics in SCADA safety, particularly for the O&G sectors, which hold potential future potential. System control is a facility in the computer-based network and technology used to inspect and control critical physical and processes operations from a distance. It gathers and analyses measurement results and operational processes in the field and transmits command signals to local or remote devices. For power structures, these control mechanisms act as a critical nervous system. A cyber assault on a management system from the outside might seriously affect public health and safety. Up until now, these control processes were thought to be secure from a cyber incident from the outside. However, remote processes in the hands of the wrong people could be used to decimate or close down network infrastructure by injecting a virus or other harmful programs into the network, causing much damage to other processes in the associated networks in the industry. Theft of data and unsupervised process design interfaces are essential issues in the O&G industry. Monitoring, screening, and updating the software we use can keep us in a secure zoo.

3.9   ROBOTIZATION AT THE DANGEROUS LOCATION FOR DRILLING FOR THE BETTERMENT AND SAFETY OF WORKFORCE

People can use teleoperation online platforms to communicate their talents and abilities to machines close by or far away (Caiza et al., 2020). Consumers may remotely access the activities and duties of machines using minimal protocols, restricting the direct involvement with dangerous locations. Because of the many chemical compounds utilized in their everyday procedures, O&G well-pads facilities are designated dangerous work zones. This trait makes these locations ideal options for teleoperation systems to avoid direct human involvement with various substances and potentially hazardous situations. The study by Caiza et al. (2020) focuses on creating a basic teleoperation method for performing monitoring and maintenance activities within one of these hydrocarbon plants. The system utilizes dispersed monitor schemes and a minimal transmission interface to manage a KUKA mobile extractor remotely to create flexible and scalable teleoperation solutions. The primary outcome of the first phase of this project focuses on creating general management and communication functions that enable a practical testing process with the aid of certified stations operators and a KUKA YouBOT mobile manipulation. Several experiments were done at Petroamazonas EP to verify the theoretical teleoperation system. The operators' adaptation to the suggested design has gone well, showing a significant improvement in their feeling of kinetics. The KUKA youBotTM robot is a fantastic addition to industrial environments since it is perfect for manipulating and moving items. The accuracy of both ending sensor placement and locomotion duties is sufficient to complete the duties assigned satisfactorily. MQTT is a communication protocol that is both quick and efficient (Siregar et al., 2020). The protocol's safety characteristics make it appropriate for use in harsh industrial environments.

O&G businesses must increase their performance, profitability, and security to meet rising energy needs (Yu et al., 2019). Yu et al. (2019) conveyed that any corrosion or fractures in their manufacturing, storing, or transport systems might result in calamities for humans and the globe. Since much O&G wealth is in harsh environments, there is a constant demand for robots to conduct inspection activities that are both cost-effective and safe. This article presents a state-of-the-art assessment of O&G examination robots, which would include unmanned aerial vehicles (Hird et al., 2017), remotely operated vehicles (Benfield et al., 2019), autonomous underwater vehicles (Okoro & Orifama, 2019), and unmanned land vehicles. Examination robots come in a variety of shapes and sizes to check various asset structures. According to the evaluation findings, trustable independent investigation of unmanned aerial vehicles and autonomous underwater vehicles will gain popularity among such robots, with reliable surroundings charting, route planning, advanced control techniques, autonomous localization, and non-destructive testing technology being the main areas of research.

3.10   PROCESS MODELING AND SIMULATION FOR THE OFFSHORE, ONSHORE AND HYDRAULIC FRACTURING BEFORE DRILLING

This research aims to present a method for developing an ML model for the O&G sector (Islamov et al., 2021). This work will focus on the most recent developments in ML and AI. One of the aims of this study was to construct a design that might identify potential hazards associated with drilling wells. Drilling wells for O&G production is a time-consuming and costly element of reservoir improvement (Mohamadian et al., 2021). As a result, in addition to hazard avoidance, there is an aim to reduce downtime and drilling machinery replacement costs. With the assistance of modern technology, firms have begun to search for ways to increase drilling efficiency and reduce the non-production period. Mohamadian et al. compared ML techniques for detecting anomalies during well boring. ML algorithms, in particular, will enable decision-making when defining the geometry of the well grid structure of the relative location of cultivation and injection wells at the manufacturing facility. The two most common types of development systems are placed along with asymmetrical grids and well placement along with non-symmetric grids. The algorithms that have been evaluated categorize drilling issues using past data from already wells drilled. Islamov et al. (2021) utilized historical drilling difficulties for 67 wells at a major brownfield in Siberia, Russia, to verify anomalous detection systems. Problematic wells were chosen and evaluated. It should be highlighted that 20 of the 67 wells were bored without incurring any costs due to wastage of time. The experimental findings show that a system that relies on gradient boosting can effectively characterize drilling operation difficulties more than other systems.

Barbosa et al. (2019) provide a comprehensive overview of the literature on ROP prediction, particularly using ML approaches, and how these methods may be utilized to improve drilling operations. Conventional methods (built on physics models) and ML methods are used to classify ROP models. This study shows that ML approaches might surpass conventional or statistical models based on ROP predictive performance. Throughout this paper, an in-depth examination of several techniques for gathering ROP models is conducted, culminating in a review of different methodologies used in the research to undertake data-driven modeling improvement. According to a literature study, considering the cost-cutting potential of real-time enhancement based on data taken from ROP models, there is a significant scarcity of application of such approaches in the industry. To continue moving ahead in real-world applications, the petroleum sector must recognize that while no general rule exists in this domain, great and very acceptable outcomes may still be obtained by adopting the best methods highlighted in this research. Furthermore, contemporary ML methods offer potential recommendations for developing tasks in the O&G sector.

3.11   FUTURE ADVANCEMENTS AND CHALLENGES FACED CURRENTLY

Fossil fuel prices continue to climb, and fossil fuel businesses need to create new digital and strengthen systems to maximize productivity and develop their current abilities (Sircar et al., 2021). Other chemicals, such as fertilizers, pharmaceutical drugs, polymers, solvents, and insecticides, need O&G as a source of energy. Sircar et al. (2021) say that technology has a significant impact on the economy and society. The confluence of innovations that blur the borders between the real, electronic, and organic domains, such as AI, robots, and autonomous automobiles, is the "fourth industrial revolution." ML has shown promise in boosting and augmenting traditional reservoir engineering approaches across many problems. AI is seeking attention nowadays due to its robust capacity and rapid generalization speed. The ANN structure is a deep learning concept that learns and understands the concepts of data. The deep learning algorithm is used to process the vast amount of information in the O&G industry, which analyses the massive data and produces the best results. Without the help of human intervention, the best features will be discovered. Moreover, deep learning algorithms can conduct advanced operations that ML algorithms cannot. Neural networks are used to process inputs. ANNs are a powerful ML tool for solving complex issues. The ANN was used in the O&G industry to handle complex nonlinear problems that a linear connection could not solve.

Changes in AI in the O&G industry, a vital element of the energy sector, are investigated (Koroteev & Tekic, 2021). Koroteev and Tekic (2021) concentrate on the upstream of the O&G industry since it is the most capital-intensive and has the most significant number of unknowns. The most current developments in creating AI tools and highlighting their influence on speeding and limiting the risks in methods in the industry are summarized based on a study of AI application possibilities and an assessment of existing implementations. We also go through the key non-technical issues preventing the widespread use of AI in the O&G sector, such as information, people, and kinds of cooperation. We also lay out three probable scenes for how AI may evolve and alter the O&G business (Tontiwachwuthikul, 2020). Even though AI is still a new advancement in the O&G industry, some usages have proven their worth. AI may help accelerate and de-risk various commercial operations in the O&G industry, including hydrocarbon discovery, field development, and raw hydrocarbon processing. AI should reduce the cost and time of research and proactive development and construction, simultaneously increasing long-term production margins. The scalability of AI is currently being tested throughout the whole industry. It covered not just the technical but also the non-technical aspects of scalability in this article. Examining how education, corporate culture, and data accessibility accelerated and directed AI adoption in the upstream O&G industry. Relying upon this study, three probable scenarios for how AI may expand inside the O&G industry in the next 5–20 years are developed.

REFERENCES

Abdalla, R. (2018). Research issues on geovisual analytics for petroleum data management. *Advances in Science, Technology and Innovation*, 185–187. https://doi.org/10.1007/978-3-030-01440-7_43

Agwu, O. E., Akpabio, J. U., Alabi, S. B., & Dosunmu, A. (2018). Artificial intelligence techniques and their applications in drilling fluid engineering: A review. *Journal of Petroleum Science and Engineering.* https://doi.org/10.1016/j.petrol.2018.04.019

Al Dushaishi, M. F., Abbas, A. K., Alsaba, M., Abbas, H., & Dawood, J. (2021). Data-driven stuck pipe prediction and remedies. *Upstream Oil and Gas Technology, 6*, 100024. https://doi.org/10.1016/J.UPSTRE.2020.100024

Al-Subaiei, D., Al-Hamer, M., Al-Zaidan, A., Chetri, H., & Sami Nawaz, M. (2019). Intelligent digital oilfield implementation: Production optimization using north Kuwait integrated digital oil field NK KwIDF. *Society of Petroleum Engineers–Abu Dhabi International Petroleum Exhibition and Conference 2019, ADIP 2019.* https://doi.org/10.2118/197811-MS

Ali, M. M. M., Zhao, H., Li, Z., & Ayoub, A. A. T. (2020). A review about radioactivity in TENORMs of produced water waste from petroleum industry and its environmental and health effects. *IOP Conference Series: Earth and Environmental Science, 467*(1), 012120. https://doi.org/10.1088/1755-1315/467/1/012120

Baaziz, A., & Quoniam, L. (2013). How to use Big Data technologies to optimize operations in upstream petroleum industry. *International Journal of Innovation, 1*(1), 19–25. https://doi.org/10.5585/IJI.V1I1.4

Balaji, K., Rabiei, M., Suicmez, V., Canbaz, C. H., Agharzeyva, Z., Tek, S., Bulut, U., & Temizel, C. (2018). Status of data-driven methods and their applications in oil and gas industry. *Society of Petroleum Engineers: SPE Europec Featured at 80th EAGE Conference and Exhibition 2018.* https://doi.org/10.2118/190812-MS

Barbosa, L. F. F. M., Nascimento, A., Mathias, M. H., & de Carvalho, J. A. (2019). ML methods applied to drilling rate of penetration prediction and optimization - A review. *Journal of Petroleum Science and Engineering, 183*(August), 106332. https://doi.org/10.1016/j.petrol.2019.106332

Benfield, M. C., Kupchik, M. J., Palandro, D. A., Dupont, J. M., Blake, J. A., & Winchell, P. (2019). Documenting deepwater habitat utilization by fishes and invertebrates associated with Lophelia pertusa on a petroleum platform on the outer continental shelf of the Gulf of Mexico using a remotely operated vehicle. *Deep Sea Research Part I: Oceanographic Research Papers, 149*, 103045. https://doi.org/10.1016/J.DSR.2019.05.005

Caiza, G., Garcia, C. A., Naranjo, J. E., & Garcia, M. V. (2020). Flexible robotic teleoperation architecture for intelligent oil fields. *Heliyon, 6*(4), e03833. https://doi.org/10.1016/J.HELIYON.2020.E03833

Crompton, J. (2015). The digital oil field hype curve: A current assessment the oil and gas industry's digital oil field program. *Paper Presented at the SPE Digital Energy Conference and Exhibition.*

Fioravanti, C. C. B., Centeno, T. M., & De Biase Da Silva Delgado, M. R. (2019). A deep artificial immune system to detect weld defects in DWDI radiographic images of petroleum pipes. *IEEE Access, 7*, 180947–180964. https://doi.org/10.1109/ACCESS.2019.2959810

Hassan, M., Rossi, D., & Ivanova, G. (2019). Lecture 1. Introduction from digital oilleld to operational excellence. *Journal of the Japanese Association for Petroleum Technology, 84*(6), 394402.

Hird, J. N., Montaghi, A., McDermid, G. J., Kariyeva, J., Moorman, B. J., Nielsen, S. E., & McIntosh, A. C. S. (2017). Use of unmanned aerial vehicles for monitoring recovery of forest vegetation on petroleum well sites. *Remote Sensing, 9*, (5), 413. https://doi.org/10.3390/RS9050413

Islamov, S., Grigoriev, A., Beloglazov, I., Savchenkov, S., & Gudmestad, O. T. (2021). Research risk factors in monitoring well drilling—A case study using ML methods. *Symmetry, 13*(7). https://doi.org/10.3390/sym13071293

Jabar, H. (2021). *Value of Digitizing Well Interventions and its Impact on Business and Working Processes [uis].* https://uis.brage.unit.no/uis-xmlui/handle/11250/2787151

Johnson, J. S., Friend, S. B., & Lee, H. S. (2017). Big Data Facilitation, Utilization, and Monetization: Exploring the 3Vs in a New Product Development Process. *Journal of Product Innovation Management, 34*(5), 640–658. https://doi.org/10.1111/JPIM.12397

Khan, J. A., Irfan, M., Irawan, S., Yao, F. K., Rahaman, M. S. A., Shahari, A. R., Glowacz, A., & Zeb, N. (2020). Comparison of ML classifiers for accurate prediction of real-time stuck pipe incidents. *Energies, 13*(14), 3683. https://doi.org/10.3390/EN13143683

Kontorovich, A. E., Eder, L. V., & Filimonova, I. V. (2017). Paradigm oil and gas complex of Russia at the present stage. *IOP Conference Series: Earth and Environmental Science, 84*(1), 012010. https://doi.org/10.1088/1755-1315/84/1/012010

Koroteev, D., & Tekic, Z. (2021). Artificial intelligence in oil and gas upstream: Trends, challenges, and scenarios for the future. *Energy and AI, 3,* 100041. https://doi.org/10.1016/J.EGYAI.2020.100041

Kshirsagar, A. (2018). Bio-remediation: Use of nature in a technical way to fight pollution in the long run. *ResearchGate.* https://doi.org/10.13140/RG.2.2.26906.70088

Kshirsagar, A., & Shah, M. (2021). Anatomized study of security solutions for multimedia: deep learning-enabled authentication, cryptography and information hiding. *Advanced Security Solutions for Multimedia.* https://doi.org/10.1088/978-0-7503-3735-9CH7

Kulyasov, N. S., Shipkova, O. T., Zavialov, A. E., & Charyyarova, G. D. (2020). Directions of development of digitalization of the oil industry in the Russian Federation. *IOP Conference Series: Materials Science and Engineering, 919,* 062032. https://doi.org/10.1088/1757-899X/919/6/062032

Lun, C. H., Hewitt, T., & Hou, S. (2021). Extracting Knowledge with NLP from Massive Geological Documents. *82nd EAGE Annual Conference & Exhibition, 2021*(1), 1–5. https://doi.org/10.3997/2214-4609.202112807

Maksimov, E. A., Vasil'ev, V. I., Soldatov, A. I., Shkerin, S. A., & Ovchinnikov, G. V. (2017). New technical solutions for the purification of oil-containing wastewater. *Chemical and Petroleum Engineering, 53*(5), 336–339. https://doi.org/10.1007/S10556-017-0344-4

Mohamadian, N., Ghorbani, H., Wood, D. A., Mehrad, M., Davoodi, S., Rashidi, S., Soleimanian, A., & Shahvand, A. K. (2021). A geomechanical approach to casing collapse prediction in oil and gas wells aided by ML. *Journal of Petroleum Science and Engineering, 196,* 107811. https://doi.org/10.1016/J.PETROL.2020.107811

Mohammadpoor, M., & Torabi, F. (2020). Big Data analytics in oil and gas industry: An emerging trend. *Petroleum, 6*(4), 321–328. https://doi.org/10.1016/j.petlm.2018.11.001

NianTian, L., Dong, Z., Kai, Z., ShouJin, W., Chao, F., JianBin, Z., Chong, Z., NianTian, L., Dong, Z., Kai, Z., ShouJin, W., Chao, F., JianBin, Z., & Chong, Z. (2018). Predicting distribution of hydrocarbon reservoirs with seismic data based on learning of the small-sample convolution neural network. *Chinese Journal of Geophysics, 61*(10), 4110–4125. https://doi.org/10.6038/CJG2018J0775

Okoro, H., & Orifama, D. G. (2019). Robotization of operations in the petroleum industry. *International Journal of Industrial and Manufacturing Systems Engineering, 4*(5), 53. https://doi.org/10.11648/J.IJIMSE.20190405.11

Orazalin, N., Mahmood, M., & Narbaev, T. (2019). The impact of sustainability performance indicators on financial stability: Evidence from the Russian oil and gas industry. *Environmental Science and Pollution Research, 26*(8), 8157–8168. https://doi.org/10.1007/S11356-019-04325-9

Orrù, P. F., Zoccheddu, A., Sassu, L., Mattia, C., Cozza, R., & Arena, S. (2020). ML approach using MLP and SVM algorithms for the fault prediction of a centrifugal pump in the oil and gas industry. *Sustainability*, *12*(11), 4776. https://doi.org/10.3390/SU12114776

Prajapati, M., & Patel, S. (2021). A review on big data with data mining. *Lecture Notes on Data Engineering and Communications Technologies*, *52*, 155–160. https://doi.org/10.1007/978-981-15-4474-3_17

Priyanka, E. B., Maheswari, C., & Thangavel, S. (2021). A smart-integrated IoT module for intelligent transportation in oil industry. *International Journal of Numerical Modelling: Electronic Networks, Devices and Fields*, *34*(3), e2731. https://doi.org/10.1002/JNM.2731

Priyanka, E. B., Thangavel, S., & Gao, X. Z. (2021). Review analysis on cloud computing based smart grid technology in the oil pipeline sensor network system. *Petroleum Research*, *6*(1), 77–90. https://doi.org/10.1016/J.PTLRS.2020.10.001

Priyanka, E. B., Thangavel, S., Gao, X.-Z., & Sivakumar, N. S. (2021). Digital twin for oil pipeline risk estimation using prognostic and ML techniques. *Journal of Industrial Information Integration*, 100272. https://doi.org/10.1016/J.JII.2021.100272

Rahman, F., Ridho, I. I., Muflih, M., Pratama, S., Raharjo, M. R., & Windarto, A. P. (2020). Application of data mining technique using K-Medoids in the case of export of crude petroleum materials to the destination country. *IOP Conference Series: Materials Science and Engineering*, *835*(1), 012058. https://doi.org/10.1088/1757-899X/835/1/012058

Rinaldi, G., Thies, P. R., & Johanning, L. (2021). Current status and future trends in the operation and maintenance of offshore wind turbines: A review. *Energies*, *14*(9), 2484. https://doi.org/10.3390/EN14092484

Rubani, M. (2017). A study of derivative market in India. *International Journal of Business Administration and Management*, *7*(1). http://www.ripublication.com

Sabri, H. A. R., Rahim, A. R. A., Yew, W. K., & Ismail, S. (2015). Project management in oil and gas industry: A review. *Proceedings of the 26th International Business Information Management Association Conference–Innovation Management and Sustainable Economic Competitive Advantage: From Regional Development to Global Growth, IBIMA 2015*, 1823–1832. https://www.researchgate.net/publication/321224310_Project_Management_in_Oil_and_Gas_Industry_A_Review

Saputelli, L. (2016). Technology focus: Petroleum data analytics. *Journal of Petroleum Technology*, *68*(10), 66–66. https://doi.org/10.2118/1016-0066-JPT

Saputelli, L. (2017). Technology focus: Petroleum data analytics (October 2017). *Journal of Petroleum Technology*, *69*(10), 87–87. https://doi.org/10.2118/1017-0087-JPT

Senum, P. D. (2021). *Market-Analysis of Software for Predictive Maintenance Solutions in the Norwegian Petroleum Industry [uis]*. https://uis.brage.unit.no/uis-xmlui/handle/11250/2786211

Shadizadeh, S. R., Karimi, F., & Zoveidavianpoor, M. (2010). Drilling stuck pipe prediction in Iranian oil fields: An artificial neural network approach. *Iranian Journal of Chemical Engineering*, *7*(4).

Shinkevich, A. I., Baygildin, D. R., & Vodolazhskaya, E. L. (2020). Management of a sustainable development of the oil and gas sector in the context of digitalization. *Journal of Environmental Treatment Techniques*, *2020*(2), 639–645. http://www.jett.dormaj.com

Sircar, A., Yadav, K., Rayavarapu, K., Bist, N., & Oza, H. (2021). Application of ML and artificial intelligence in oil and gas industry. *Petroleum Research*. https://doi.org/10.1016/J.PTLRS.2021.05.009

Siregar, B., Pratama, H. F., & Jaya, I. (2020). LPG leak detection system using MQTT protocol on LoRa communication module. *2020 4th International Conference on Electrical, Telecommunication and Computer Engineering, ELTICOM 2020- Proceedings*, 215–218. https://doi.org/10.1109/ELTICOM50775.2020.9230518

Solanki, P., Baldaniya, D., Jogani, D., Chaudhary, B., Shah, M., & Kshirsagar, A. (2021). Artificial intelligence: New age of transformation in petroleum upstream. *Petroleum Research*. https://doi.org/10.1016/J.PTLRS.2021.07.002

Tontiwachwuthikul, P. (2020). *Recent Progress and New Developments of Applications of Artificial Intelligence (AI), Knowledge-Based Systems (KBS), and ML (ML) in the Petroleum Industry.* https://doi.org/10.1016/j.petlm.2020.08.001

Vijay, A., & Unni, V. S. (2012). Protection of petroleum industry from hackers by monitoring and controlling SCADA system. *All Days.* https://doi.org/10.2118/149015-MS

Wanasinghe, T. R., Gosine, R. G., James, L. A., Mann, G. K. I., De Silva, O., & Warrian, P. J. (2020). The internet of things in the oil and gas industry: A systematic review. *IEEE Internet of Things Journal*, 7(9), 8654–8673. https://doi.org/10.1109/JIOT.2020.2995617

Wang, L., Li, J., Peng, K., Jin, S., & Li, Z. (2004). Petroleum pipe leakage detection and location embeded in SCADA. *2004 International Pipeline Conference*, Volumes 1, 2, and 3. https://doi.org/10.1115/IPC2004-0717

Yu, L., Yang, E., Ren, P., Luo, C., Dobie, G., Gu, D., & Yan, X. (2019). Inspection robots in oil and gas industry: A review of current solutions and future trends. *2019 25th International Conference on Automation and Computing (ICAC).* https://doi.org/10.23919/IConAC.2019.8895089

Chapter 4

Midstream sector with ML models and techniques

4.1 INTRODUCTION TO MIDSTREAM SECTOR AND ADVANCEMENT WITH THE ML TECHNIQUES

The midstream sector of the oil and gas industry is the middle segment between the upstream and downstream or in simple words the sector between the production and exploration of oil and natural gas to the processing and converting them to end products for the customer. It majorly incorporates the storing, transporting, and marketing of oil and natural gas. Majorly the companies functioning fall under a category which does all the three things or sectors altogether, i.e., They operate in the upstream, midstream, and downstream sectors because their activities and expertise in the work have always been a part of their operations. They also operate in the midstream and downstream domains.

As the largest privately owned pipelines and storage facilities in their respective areas, the United States and Canada are becoming more prominent as the major epicenters of midstream-based industries, for example, TransCanada Corporation, Oasis Midstream Partners, Sanchez Midstream Partners, and EQT Midstream Partners. If we talk about the Indian market, Indian Oil Corporation, Hindustan Petroleum Corporation Limited, Adani Enterprises Limited, and Gail Limited are found to be the main leaders.

With the vibrant panorama for energy production, artificial intelligence (AI) and machine learning (ML) deliver powerful and influential benefits and assistance across the all-inclusive value chain. AI and ML help oil and gas utilities, and the research and development assess and appraise the value of the specific reservoirs, customized drilling, and completion plans according to the geology for the particular area, and assess risks of each well and the utilities. Major domains we can see where AI and ML tools help in the midstream sectors are field processing, storage, and transportation.

Conventional computer systems necessitate a present and pre-planned program to do or perform any task. No concern with the complexity of a task it can perform that task only not beyond that. ML on the other hand is a computer system that can acquire to perform automated computerized tasks and think for itself through the algorithms and data provided to it along with absorbing new data sets and experiences.

ML initiates with the basic algorithms and evaluates and analyzes a large data set and then makes the predictions depending upon what it finds in the data and current updates. The algorithms apply that knowledge and information to learn new and innovative models and techniques for analyzing and acting upon the loopholes and

DOI: 10.1201/9781003279532-4

bottlenecks and acting for forthcoming data. It is the current and future technology for automation tasks and its pre-requisite construing of data sets and making the predictions with speed and efficiency.

The ordinary oil and gas business is not able to design robots that use AI, and they are utilizing equipment that can be programmed and customized for performing routine tasks. Furthermore, those daily tasks often require analyzing complex data sets so that work is maximized and efficiency and effectiveness and return on the investment are improved.

ML has an astonishing perspective for the ever-changing game in the oil and gas industry comprising the following domains:

a. Automation.
b. Collection of data and assessment of data.
c. Complex algorithms.
d. Analytics in an expendable format.
e. Recommendations for the bottlenecks and improvements.
f. Maximized and accurate efficiencies.
g. Automated alterations and modifications.

The midstream facility is one of the more unpredicted aspects of ML in the oil and gas industry in the journey from fields to refineries analyzing gathering, transportations, logistics, and pipelines.

Algorithms can crunch numbers so quickly that they can provide specific suggestions and recommendations for improving the effectiveness of your systems.

Thus, creating such ML tools and programs can help in the operations, monitor the operations, and help in the functioning and operating. The AI and ML combination can be utilized for better operations and monitoring (Figure 4.1).

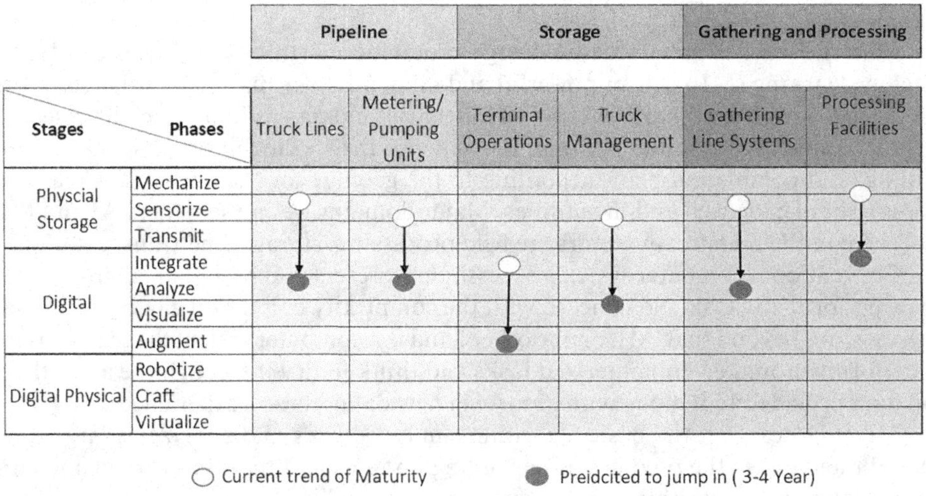

Stages	Phases	Pipeline		Storage		Gathering and Processing	
		Truck Lines	Metering/ Pumping Units	Terminal Operations	Truck Management	Gathering Line Systems	Processing Facilities
Physcial Storage	Mechanize					O	O
	Sensorize	O	O	O			
	Transmit						
Digital	Integrate			O			●
	Analyze	●	●			●	
	Visualize				●		
	Augment			●			
Digital Physical	Robotize						
	Craft						
	Virtualize						

○ Current trend of Maturity   ● Preidcited to jump in (3-4 Year)

Figure 4.1 Digital maturity of midstream operations.

Like if we see the trend for the pipelines section the trend is for the physical to the digital. For the storage domain, it's from the digital to the digital–physical systems, and for the gathering and the processing domain, it is from the physical to the digital transmission.

4.2   TRANSPORTATION WITH THE PIPELINE AND THE DIGITAL MONITORING SYSTEM

Transportation of oil and gas via pipelines is the largest mode of transportation with more complexity and other external conditions. Pipelines are extensively used for the transportation of hydrocarbon fluid over squillions of kilometers all over the world. The constructions of the pipelines are designed to endure numerous environmental lading surroundings to confirm safe and consistent delivery distribution from the source or the distribution depot. A keynote aspect for the sustainable and maintainable development of the oil and gas industry is the remote monitoring of integrity and reliability of transportation pipelines. There are at present many technologies established for monitoring and giving accurate results. The task for the next era is to integrate the established technology and drive the entire operations based on the data created by the existing systems till now. In ML- or AI-based technology, the key element is the data. In the past, there was a problem in communication. But we have surpassed it. Now that communications and sensor technology are established, we need to integrate with the machine and deep learning for effective and accurate outcomes. Algorithms and models should be developed in such a way that we don't need to establish a new green root or start from the initials. Integrated sensor networks and pressure pulse receivers are established technology in some of the networks of the pipelines. We need to connect that technology with the ML platform with the algorithm so that we can have one digital monitoring system for transportation. The advantage of the one system monitoring will be that the operation and the analysis become comparatively easier than the previous ones. Engineers operating can upgrade the operations smoothly and have a unique control under one tree. The challenge is to develop algorithms and the code for such complex pipe networks and have prediction accuracy of that. But seeking the potential of the researchers, engineers, and research and development seem too doable and we can make the system automatically learn from the data and increase its capacity. This is the beauty of ML in that it learns itself as per the data concerning time increases and sets things in such a way. The need for developing such an application or technology is to monitor pipelines to see if there are any leaks, and if there are, we can immediately stop the flow from that particular technology on the spot and repair it and replacements and the operations can restart as well as another benefit I can see is that calculations can be done for further storage, selling, refining process accurately as per the flow and the storage facility available so that efficient business can be done without any obstacles of not having the information.

Olugboji et al. (2021) presented the study of an Internet of Things (IoT) analytics platform facility to impersonate real-time pipeline monitoring systems and estimate the location of the damage or leakage on a pipeline. They used the pressure pulse based on the norm of pipeline vibration to analyze the working. They allotted the principle of time delay between pulse arrivals at the spot of sensors. A wifi-module and an

Arduino-module setups were combined, programmed, and used to develop a wireless communication mechanism that communicates through the ThingSpeak IoT platform for analyses. The developed transmission device endorsed the ability to analyze via testing it through test rings with static air. Based on the earlier developed methodology of locating an interruption event on a pipe that depends on the delay of the time pulse influxes at two sensors, an alteration of 20 mm approximate was recorded between the computed and actual event locations when a data logger was employed for apprehending and communication of sensor data. They carried out this experiment via validation of the pipeline monitoring theory based on the arrival of pulse times at sensors along the pipe. Thus, they incorporated the IoT platform and managed the network. We can incorporate this technology in our one tree of ML program and track it directly in our system and have visible 3D monitoring if the advanced sensors are used. The initial cost of the sensors would be high if there is no installment of any such sensors in the pipeline. But if there is one installed, we can upgrade the program and the coding and can achieve our target.

Inline to mitigate the risk associated with the third-party interference risks and to reduce the environmental exposure, it is possible to organize multipoint acoustic sensing technology in which the multiple sensors were located at a fixed distinct distance along the pipeline (Giudice & Milano 2014). Any interactions generated in the flow in the pipeline, acoustic waves are generated and directed within the fluids (gas, oil, and any hydrates) if present giving the information regarding the source or any flotation in the flow regime. These wave propagations are directed by the velocity of the sound and coefficient of absorption which are the functions of the pipe or we can say depends on the nature and the material of the pipe and the nearby medium. These characteristics have been examined by administering real-time data amassed with an exclusive Multi-Accounting System (e-vpms™) on the fluid transportation in the various flow conditions and the on-time present operational conditions, generating inclusive data sets and third-party interactions, Trials for the leaking and the inspections of the pig.

The technology seems to give promising results for real-time monitoring along with observing and censoring the advanced intervention planning. The by-product we obtain, i.e., data collected from the sensors can be utilized for smooth operations and predictive things for the future as well as for designing the system or advanced system.

As a result, we can use this technology if we have the budget for it and where third-party interactions are frequent and time-sensitive.

We can amalgamate it with our ML tree concept where we can set the algorithm in such a way that the 3D printing concepts can be utilized and make the similar kind of broken or unmatched parts and don't have to wait for cutting and making the pipes to start the system as soon as possible.

Efficiency, safety, and sustainability are the significant domains of the modern oil and gas industry and in the subsegment of the pipeline transportation systems too. Pumps play a very vital role in the transportation of crude oil and the pumping system for the system of transportation of oil and gas pipelines. Along with monitoring of the pipelines and the leakages and having the constant flow we need also to take care of the pumping segment under our great tree of ML. Monitoring all the assets and having control over the operation is a tricky as well as challenging task. But where there are challenges there are opportunities too. Let's turn this crucial task into a safer and sustainable reference. Sensors are applied to the pumps for operation, but in some cases, in old machinery, it is not possible to run them via remote operations so in that case alternate arrangements should be made in such a way that the working network is not disturbed and the and the pup can be operated remotely via considering the existing alliance. Angelo et al. (2021) mentioned in their work a predictive maintenance technique where the condition of a centrifugal pump and the other utilities is tracked remotely and develop such a model which can be operated also with the help and on the platform of the IoT and ML. The pressure can also be predicted from not exactly a proven approach but trial and error methods. The smart monitoring technology was offered and authenticated on conventional pressure signals collected by Eni for numerous years on crude oil transportation pipelines located in similar operating conditions. Hence we can also predict the maintenance, operating conditions, and real-time conditions of the pumps and also try to operate them remotely through ML techniques.

4.3 OPTIMIZING PIPELINE SCHEDULING FOR PRODUCT FLOWS

As we are moving toward digitalization and automated operations, the pipeline scheduling and product flow play an important role in the transmission to verify the leakages, bends, disturbances, and uninterrupted flow. There are quite numerous incidents like nodes, vandalization, leakage, and corrosion being the barriers for the pipeline's transportation of the crude oil in almost every country as they are dependent on the surrounding conditions and the quality of the pipelines as well as on many other things. These failures have resulted in annual losses of up to around US$10 million in the United States (Slaughter et al., 2015). Environmental casualties are another consequence (Ambituuni et al., 2014) and deadly incidents, which documented the deaths of thousands of people in the Jesse in 1998.

(Ambituuni et al., 2014). Subsequently, many techniques are arranged and installed to monitor and optimize the people and the flows for reducing the mitigation failures and other losses. Currently monitoring systems include daily overflights using high-resolution cameras by some special helicopters, drones, and local surveillance methodologies. The main thing commonly in them is all based on human interventions, and they are diagnosed at a particular time in their presence, not any real time-based. Also, they are high-priced machines. Unconventional wired systems utilizing fiber optics or copper-based methods are used but are limited by abnormal capital installation costs and operational as well as maintenance costs. Also, there are more complications in fault location and repairing, non-flexibility to thrash terrain, and the possibility of the third-party interaction (Aalsalem et al., 2018).

So, there is a need to establish cost-effective, easily applicable, and more specific and finest and elite algorithms and coding using ML for better operations and monitoring things. It's a fact that we all are moving toward the renewable energy sector and crude, and oil is getting less demand. But it's too a fact that crude is going to serve us and the drastic change or transformation toward the renewable sector will take time for establishing. The benefit of doing the digitalization and ML-based operations and the unit process is the optimum usage of the crude we have and are generating. The optimum usage will reduce the wastage, as well as enhance the operation which will further save time. Once the algorithms and the programs are created they can be employed in many other applications for monitoring such flows with making certain changes. Base coding and digitalization are more important to us rather than industry. The industry is important but we can change or allot to other sections, tailoring and making it suitable for the industry. With the new cutting-edge technology, we are in Industrial Revolution 4.0.

Ahmed et al. (2020) presented the IoT-based monitoring system for pipelines for oil and gas pipelines ensuring and validating the resiliency across the network complex. They proposed the system being an ascendable, efficient, and cost-effective model by taking into account trade uncertainties and the fluctuations of the property such as density of the fluid, detectability, and expenditure. The application and the operation were simulated and evaluated for the performance of the system. The primary results for the consolidated detection or localization demonstrate the performance for each methodology, i.e., detection or localization was calculated and resulted in it.

They used data replication and asynchronous consensus across the nodes for scattered and circulated decisions with the hope of an IoT- and ML-based platform.

In addition, they implemented features for data scanning and filtering, categorizing and grouping, prioritizing and ordering historical and currently generated data computations for trends and emergencies, and understanding the behavior of systems with changes in parameters.

Further work is required with the modeling and the algorithms (Figure 4.2).

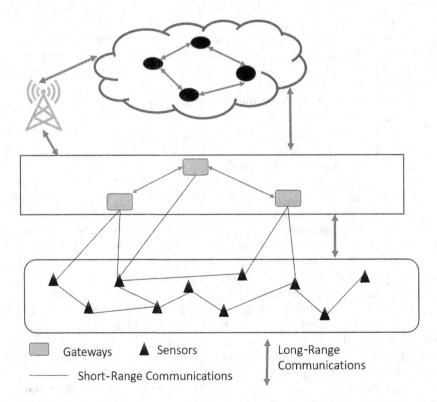

Figure 4.2 This is in reference to the work of placement and Distributed Estimation for pipeline leakage detection and localization (Ahmed et al., 2020).

Adegboye et al. (2019) in their paper gave guidelines for the readers and the developers for the selection of suitable technology for detecting the leakages for particular locations and settings. They discussed the work done in the field and the need for more work being required and the open issues and the gaps between the model and the actual practical situation. They demonstrated the exterior methods and the designing of the sensing systems and their monitoring of the external systems of the pipelines. Acoustic emissions sensors, fiber optics sensors, vapor sampling, ground-penetrating agreement radar agreement, and infrared thermography were among the methods included in this category.

The second category was the visual techniques for detecting the leakages of the pipelines. They include well-trained dogs as well as field experts with years of experience. Smart pigging techniques, as well as helicopters/drones/remotely operated vehicle (ROVs)/autonomous underwater vehicle (AUVs).

The internal method of detection of leakage and its properties are associated with the hydrocarbon fluids like flow rate balances (mass or volume). Negative pressure waves, pressure point analysis, digital sensing processing, and dynamic simulations were integral parts of the third category. Then they comparatively carried out the analysis given by the American Petroleum Institute guidelines. Based on their analysis, they found that each method has some pros and cons, but the third category shows some positive prospects regarding it. We can employ the ML tools and the IoT-based platforms for the data gathering and design of the model incorporating all the drawbacks of the one model and the technique for operating and monitoring (Figure 4.3).

In the undersea production systems, the reservoir contains a mixture of various compounds. When the temperature and pressure conditions in the system or surrounding the resins change, the formation of various types of paraffin and hydrates occurs, and they are deposited in pipelines, obstructing the flow. Flow assurance is the concept and engineering methodology used to ensure the smooth flow of hydrocarbon fluids and the project's economic life in the worst environmental conditions.

In general, depositions take place. Reducing or cutting down the production of resins or hydrates deep down the sea is harder, and it also decreases the cross-sectional area of the flow in the pipe and disturbs the stability of the flow. Gas hydrates are generally created in gas transmission and production. These gas hydrates are ice-like solid compounds generated under low-temperature and high-pressure conditions and act as an annoyance in the gas pipeline (Menon et al., 2005). Depressurization and injection of the inhibitor are the solutions for that but they are found to be costlier and time-consuming. Annually, hundreds of millions of US dollars were devoted to preventing hydrate formation. The system for the diagonalization should be adapted with ML and AI for better accuracy over the existing mathematical models. To determine the model, ML is obligatory. However, these hydrate's formation takes place under seabeds so collecting the data over here was practically not possible but applying the concept and the model of the digital twin. This was the combination of physical and waterproof-based sensing technology with cloud computing and the IoT-based application.

Seo et al. (2021) in their study concluded that the model hyper-parameters matched and structurally established the optimization with the SAE model and the greedy layer-wise methodology. They concluded that with the time-series forecast the model was able to detect the growth of the hydrate volume. The accuracy was related to the

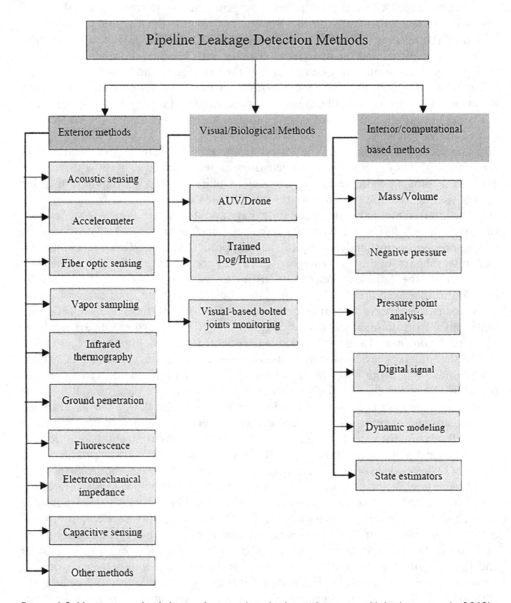

Figure 4.3 Various methodologies for pipeline leakage detection (Adegboye et al., 2019).

volume of the hydrate. The more the volume of hydrate, the more the accuracy of prediction. The actual data were matching the trends for the growth. The overall performance was validated via average R-squared for the volume of the maximum hydrate to be 97% and that for the formation was found to be 99%. Thus, the combination of AI and ML can be applied for predicting the growth of the hydrates in the pipelines. Further enhancing the technology we can use the data and operating conditions the model can be developed in such a way that it can accurately find the location and the size of it.

4.4 IMPROVING RELIABILITY RISK MODELING FOR REFINING AND PROCESSING ASSETS

Managing the risks is a process that aims to decrease the hazardous effects of the actions via constant activity or motions of the fluid to foresee and forestall the undesirable affairs or the events happening and the strategic planning to circumvent them. Risk management can occur via a process of determining the risks and then tailoring the designing the model and the strategies for managing the risk. There are some major incidents in the industries due to improper operation of the assets and the types of machinery. The impact is on the safety of the workers, environmental impacts, and economic loss to the system and the utilities functioning there.

The failure or the incident not only wastes the time for the operation to regain but the reputation and the image in the globe of the company also decreases. So, risk-based reliability models and technology should be developed by the company with the current and future technology, such as AI and ML. This technology is not as mature so they are found costly. But with time and research and development, they will be commercialized and will be economically viable.

Proper assessment methodology and risk-based maintenance reduce the consequences of asset failure or human error. One of the main engineering glitches in risk modeling is the uncertainty and the unavailability of the information. Information is then converted to the data and these are the basic prerequisites for ML. The first stage of technology development is the generation of data in all forms, 2D, 3D, information, specifications, vibrations, pressure drops, sensor data, drones' images every information linked with the operation must be considered as data. These sensors and the ML tools and the programming may be considered costlier reason being the fundamentals are not ready yet, the proper directions are not available for the coders to build the algorithm, this particular manuscript will give directions for future computer science-based engineers and the data science engineers for building the complex algorithms and the costlier technology to the economical and make the nation's own proud and the business strategies. Every nation should possess its methodology and the platform to reduce its dependency on other countries.

The value chain and the efficiency of the petroleum industry depend upon many essential factors and many factors play a vital role. For improving the performance and reliability the pipelines are taken for the investigation and play a significant role. Visual inspection and a detailed survey of the utilities in operation and the facilities available resulted in a legitimately reliability evaluation of the conditions.

It requires a large amount of time for the collection of useful data for utilizing the massive facilities. Luis and Gonzalez (2016) proposed a model intended for giving an integrated risk assessment solution for the terminals or the end-stage where there is a combination of the process and the functions with the operations as these things should have proper safety with a mutual purpose. A supporting tool was the software railing on the operator's and the field engineer's approval according to their requirement. They applied the current model to an SDT situated in Northern Mexico. The software being utilized covers the stations, terminals, and ducts the advantage is that it is applied directly without being given training. The requirements were the GIS and the management of the database and the data too. They aim to apply to other utilities and get the benefit of it. The algorithms can also be developed along similar lines with the help of GIS and location-based mapping for the desired outcomes.

Sa et al. (2014). The conventional and fuzzy-based risk assessment method for the modeling of process operations in the petroleum industry is described as risk-proof in their work.

The model was based on the real-time condition of the operations. The accuracy was a function of accurate data and the precision to ensure safety throughout the processing facility and monitoring and detailing the failures. The fuzzy-based models were found to be more precise than conventional RBM ones in the application in the gas plant study. The fuzzy method was promoted because it can work with inaccuracies in price data or information when there is uncertainty. A similar kind of RBM fuzzy model can be useful to other industries dealing with the risks of the process and the operation. They proposed the model and believed that it is a benchmark for future work.

The presented model was unable to produce an overall unified ranking; it requires additional information for assessing the robustness of the original or novel scales when the uncertainty is there. A similar kind or the basic needs for that can be utilized for it.

4.5   IMPROVED STORAGE AND PROCESSING FACILITIES

The storage of oil and natural gas supports the easy and persuasive demand irregularities also the pricing system too. Generally, it is observed that utilities store the oils and gas when the process is lower relatively and realize the stock when there is a hike in the demand and the prices evaluated.

In the Covid-19 pandemic, it was observed a drastic decrease in the oil demand which led to the oversupply of oil due to Saudi's hiked oil production, and countries OPEC and non-OPEC countries failed to come to an agreement of lowering the production (Brown et al., 2020). The storage tanks were at the peak of their capacity. Responding to that, oil storage utilities raised the rates of storage. For example, tankers were charging around $24,000–$25,000 per day in February 2020, but that was increased to $295,000–$300,000 per day in April 2020.

But it was the conditions and the period which was not in our control so we needed to go with the flow. But to avoid such similar situations in the future we need to develop the technology for that. This particular content is reliable for the government and the managing and regularities authorities of the country. For ensuring smooth operations, as well as digital monitoring systems for storage and additional processing facilities.

Companies have built the assets and invested in the storage facilities so they are going to take advantage of them.

In the storage chain, there are insufficient supplies and demand irregularities; sometimes there is no space for additional storage, and sometimes there is space available for storage, but exploration is ongoing.

Also, there are cases wherein some utilities are fully working and don't have the additional storage system, while in some cases due to external factors such as locations, cost, and the facilities provided may not be suitable to someone and they might not be working as effectively. To overcome all these uncertainties and irregularities, we proposed integrated solutions and a platform using the new leading-edge technology AI and ML. A digital platform provides all relative information to the government or the regulating authorities for giving regular control in the prices to stop the jerk of the demand rises and avoid the utilities being remaining empty. The platform will give

ML-based and AI. The system will be integrated and connected with all the storage facilities available in the country. This proposed system is for national authorities in charge of oil supply and energy needs, and it can be implemented by them.

The processing facilities are generally the refineries where the crude oil is being fractionated into useful products. The biggest names in the world domain for the production of oil and gas are Saudi Aramco having a capacity of 10,963,091 barrels of oil per day, Rosneft having 4,217,780 barrels per day, NIOC, CNPC, ExxonMobil, Petrobras, etc. Each refinery is designed according to the raw material, processing, or refining capacity and the outcomes of the maximum achievable production. The operations are carried out in a semi-automatic manner. The system has been set up using the PLC and SCADA systems for the operations but to set up the program and to operate the process human interventions are necessary. Generally, chemical engineers are employed for the task as they are the most suitable and applicable domain that can handle such complex processes and operations. Chemical engineers are familiar with the processes carried out, their particular operating conditions of the particular distillation or cracking reactions, and other things carried out. To assist them and make their work more efficient we can have the AI- and ML-based software and the computer-aided tools ruined by the data generated from the past decades of the same particular process. We can have the prediction timings and conditions and the special requirements needed for a particular oil. In steam or hydro-cracking, operating conditions such as pressure, temperature, and temperature of the steam are used, as well as other important aspects.

The shutting and the rest time and all the other information can be calculated using the ML approach and with the help of AI, we can have the connectivity between the engineers and the operating office. As there are issues of data privacy nowadays, companies generally restrict the data and other important things to employees. So we can provide them with the main results of the calculations generated by our software digitally and the 3D output of the results. The bigger speed breaker we observe is the extensive paperwork. For better communications and proofing reports, we can have digital devices, such as tablets, palmtops, and laptops for connectivity with the inbuilt walkie-talkies installed for vocal communication. The environmental concerns and the dependency are also reduced with a paper-free ecosystem with AI and ML.

4.6   MAXIMIZING LABOR PRODUCTIVITY AND WRENCH TIME VIA EMPLOYING ROBOTIZATION AND ML TECHNIQUES

Robotization snatches the job of the on-field worker but is employed for the hazardous job which is the life of the workers we are employing. It's better to employ machines rather than the life of a person in a hazardous task. To increase the productivity and the efficiency of the operations, the robotics digitalization is applied. When there are irregularities in the oil and gas industry there are also ineffectiveness and inefficiencies being seen in the operations in the production, due to that revenues are dropped which keeps the need for applying the automation. The initial installation cost is higher but overall operations can compensate and result in the overall gains. The general stage is the production phase. This includes the remotely operated aerial drones and underwater vehicles which are generally used for drilling and underwater welding. Roughneck robots are employed for the dangerous job which connects drill pipes in stimulating locations like oil-bearing rocks and the ocean. Automated rigs are the innovative design employed for safety operations. Underwater robots spend more time in the water performing maintenance and necessary repair work.

They are connected and operated from the automated rig. There is also a development with the connectivity of the ethernet 4G technology. Downtime spent on a rig and other drilling sites is costly with robotic drills.

Robots can solve this problem and hike productivity. Thus, looking for the safety of the workers and enlarging the working capacity or the productivity of the operations robotization should be adapted and connected with software-based automated remote operations. This can be done via using ML tools and algorithms and AI. We can connect and integrate the existing technology of robotization with self-operated and automated operations.

The extent of measuring the planning, scheduling to increase productivity is called wrench time. It's the in-house study and the calculations done to obtain the optimum conditions. If we say in simple words the wrench time is the time devoted by the workers, employees, and any other human resource for a particular task from their given total time is considered as wrench time. The importance is stated by Richard (2019) that the average wrench time is in the range of 25%–35%. Planning and scheduling can boost the wrench time and can be evaluated around 50%–55%. And this is stated as a world-class performance. It is an achievable target and many plants have done this. The leverage ratio is calculated at around 55 divided by 35 and this factor is multiplied by the number of workforce available.

Thus, we can develop the software-based supported by AI and ML, the automated planning and scheduling work for the on-field workforce and also for the employees based on the work pending, priorities, urgency, and the availability of the materials available. Basic planning can be done on a day-to-day basis and can be checked and verified by the employees and the necessary changes required can be done based on their expertise. This tool can realize the major burden of the planning team and the load on them. Also from the data available, it can learn itself and carry out things. In a single day, this cannot be developed; this requires the time, effort, and dedicated mindset for developing such algorithms which are more complex. Generally, data privacy issues are created for such applications, so they should ensure development and controls can happen for the data privacy as data privacy is one of the biggest issues of today's world. For creating the in-house technologies companies have to hire suitable developers and pay them for creating such things rather than buying the utilities from other sources. One can buy from big companies such as Oracle and TCS, but we don't have trust in anything; nevertheless, they are being the best and trusted and reputed service providers. The time management and the planning of day-to-day work can be done on the proposed tool.

4.7 PREDICTING THE SUBSEA AND GROUND PIPELINE BY ML TO OPTIMIZE LATERAL BUCKLING MITIGATIONS

With the increased temperature and pressure, the instability conditions are growing majorly of the lateral buckling and the axial walking of offshore pipelines and among these pipeline operators. The pipes get bent vertically on one side, and if stressed more, they can break and a huge amount of oil can be wasted, and the subsea ecological balances and biological environment can be harmed.

As with the increase in the temperature, the thermal expansion is also increasing and this is a critical design issue of the pipelines as the capacity of the expansion and the transmission are in limited sources. Axial ratcheting behavior is to be considered beyond the conventional phenomena of limited space expansion. This kind of propaganda is known as pipeline walking which generally is responsible for the overall movement of the pipes. Pipeline walking causes disturbance in the flow and connection failures. Pipeline anchors are to be installed as the mitigation way for the pipeline walking strategy. So it is necessary to have lateral bending on the shore. While there are some mitigation options available to stop the bending phenomena like some pipelines designs. But it is necessary to have a prediction over it. Pipelines global bucking is one of the diagnostic and critical processes and requires numerous manual efforts. Some of the semi-automated software is available on the local PCs and they don't have 3D monitoring. The cloud-based and fully automated system has to be provided with the all-time 3D analysis and the leakage optimization.

Bhowmik et al. (2020) presented the Digital Field Twin digital study of the subsea design of process and the pipeline study adding the disruption of the conventional process works concept phase of the operation cycle of the assets. The digital twin field is developed in agreement with open-source coding and commercialized software for the automation design of processes and pipelines. Almost all calculations of the pipelines were able to perform in it. And they are done via python programming. Their proposed digital twin fields can carry out each pipeline and structure design calculation through cloud-based connectivity with the American Petroleum Institute (API).

They proposed Abaqus and Orcaflex for the detailed FE analyzer and riser. The work efficiency and the cost of human resources are reduced by opting for automated design approaches in a digital field. In the early planning and the execution of the project if the automated design is available we can affect incorporating all the pipeline design calculations and automated report formation.

We can expand it with the ML tool and AI-based methods for calculating the break time and the strength of the structures. ML can be employed by consuming the data of such incidents and similar pipeline characteristics and operating conditions.

Subsea fixed buoyancies are one of the key options or methods available to mitigate pipeline lateral buckling caused by extreme pressure and temperature cycles caused by frequent shut-ins and restarts. These subsea pipelines are assembled with fixed buoyancies to move and expand over the many operational cycles and can be dug into the seabeds creating the mounds of the soil in the outer area. These soil-fixed berms prevent the movement or the motion of the pipes and may lead to a decrease in the life cycle and the extension of the life is also reduced. Subsea spinning buoyancy or resilience permits the pipeline to move simply over the surface of the soil. We can minimize the soil berms and stress building up and help the integrity of the pipelines (Critsinelis et al., 2020).

Digitalization and automated operations based on the data and the faults found concerning the real-time monitoring solutions are required. We can also design the subsea robots and the technology such that if the leakage or the breakdown of the pipe is found, it can fix that by going there and having a repair over there so that accurate results can be obtained.

4.8 DATA MANAGEMENT FOR EQUIPMENT AND FACILITIES ALONG WITH OPTIMIZATION AND PROCESS CONTROL WITH AUTOMATION

Data is the basic raw material for creating any ML tool or algorithm. Even the database management system is very critical for having a glance overview of the equipment in usage, maintenance, and the equipment needed and necessary for the operation. Based on the priority and the necessity we need to arrange them on the demand. Data management of such equipment and facilities is required, and if it is done using AI- and ML-based technology, we add value through innovation in our creation.

A digital optimized internal connection and the system should be developed such that the real-time entry and the current status of the facilities are known on the demand. It helps in the planning and buying or arranging of the utilities required. Data sets and their privacy are the key challenges faced and will be the bottleneck for data scientists and engineers. We have to build our empire and the digitalized operations based on the data and also protect it. MNCs like TCS and Oracle are focusing on future trends and trying to provide security to the database. Process control and optimization are vast oceans; there are numerous positive and promising segments where such technology is desperately needed to make operations faster and more accurate. Such as searching and browsing specific data files in complex systems and databases. Details about the specific utility or service provider that it manufactures or provides.

We need to develop such algorithms based on the data and the inter-exchangeable forms. Connectivity of the 3D models and each operation PC is essential. If a single station engineer adds data to the database, it should have a higher synchronizing capacity and be available to perform the work.

The complex network should be built in such a way that keeping the probability of the insisted surf or the entry, it should show the results in such a way we don't need to surf more in the complex network. Initially the semi-automated and semi-digital systems are developed and validated and then we are focusing on the overall automation. The similar results and the similar search optimization should be carried out first then we should try on the analytics. Once the basic platform is made we can enhance it by making it a cloud-based operation and the remote operation for the valid field pics.

REFERENCES

Aalsalem, M. Y., Khan, W. Z., Gharibi, Khan, M. K., & Arshad, Q. (2018). Wireless sensor networks in oil and gas industry: Recent advances, taxonomy, requirements, and open challenges. *Journal of Network and Computer Applications*, *113*, 87–97.

Adegboye, M. A., Fung, W. K., & Karnik, A. (2019). *Recent Advances in Pipeline Monitoring and Oil Leakage Detection Technologies: Principles*. Basel, Switzerland: MDPI

Ahmed, S., Le Mouël, F., & Stouls, N. (2020). Resilient IoT-based monitoring system for crude oil pipelines. *2020 7th International Conference on Internet of Things: Systems, Management and Security (IOTSMS)*., Valencia Polytechnic University, Valencia, Spain.

Ambituuni, A., Amezaga, J., & Emeseh, E. (2014). Analysis of safety and environmental regulations for downstream petroleum industry operations in Nigeria: Problems and prospects. *Environmental Development*, *9*, 43–60.

Angelo, R., Bernasconi, G., Giunta, G., & Cesari, S. (2021). Journal of petroleum science and engineering a data-driven pipeline pressure procedure for remote monitoring of centrifugal pumps. *Journal of Petroleum Science and Engineering, 205*, 108845. https://doi.org/10.1016/j.petrol.2021.108845.

Bhowmik, S., Naik, H., & Mcdermott International. (2020). *Digital Field Twin for Subsea Field Development*. Texas, U.S.: OnePetro – SPE, 4–7.

Brown, P., & Ratner, M. (2020). Low oil prices and U.S. oil producers: Policy considerations. *Congressional Research Service Insights*, IN11246, April 1, 15–25.

Critsinelis, A., Mebarkia, S., & Huang, M. (2020). *OTC-30865-MS First Deployment of Subsea Rotating Buoyancy to Mitigate Lateral Buckling of High-Temperature Pipelines Pipeline Lateral Buckling Field Observations* Texas, U.S.: One Petro – SPE.

Giudice, S. D., & Milano, P. (2014). *Advanced Real-Time and Long Term Monitoring of Transportation*, no. November. https://doi.org/10.1115/IMECE2014-36872.

Luis, J., & Gonzalez, M. (2016). *IPC2014–33331 Risk and Reliability Modeling to Support Logistics Performance*. New York, U.S.: American Society of Mechanical Engineers, 1–10.

Menon, E. S. (2005). *Gas Pipeline Hydraulics*. CRC Press: Boca Raton, FL.

Olugboji, E. N., Aba, O. A., Nasir, A., & Olutoye, M. A. (2021). Petroleum pipeline monitoring using an internet of things (IoT) platform. *SN Applied Sciences 3*(2), 1–12. https://doi.org/10.1007/s42452-021-04225-z.

Richard, D., & Palmer, P. E. (2019). Maintenance Planning and Scheduling Handbook, Fourth Edition. DIY Wrench Time Study, Quick and Easy In-House, Chapter (McGraw-Hill Education: New York, Chicago, San Francisco, Athens, London, Madrid, Mexico City, Milan, New Delhi, Singapore, Sydney, Toronto.

Sa, E., Anvaripour, B., Jaderi, F., & Nabhani, N. (2014). Fuzzy risk modeling of process operations in the oil and gas refineries. *Journal of Loss Prevention in the Process Industries*. https://doi.org/10.1016/j.jlp.2014.04.002.

Seo, Y., Kim, B., Lee, J., & Lee, Y. (2021). *Development of Ai-Based Diagnostic Model for the Prediction of Hydrate in Gas Pipeline*. Basel, Switzerland: MDPI, 1–22.

Slaughter, A., Bean, G., & Mittal, A. (2015). *Connected Barrels: Transforming Oil and Gas Strategies with the Internet of Things*. Deloitte Center for Energy Solutions, Tech. Rep.

https://ecorobotics.com/industrial/how-robots-are-helping-the-oil-gas-industry/

https://www.automate.org/blogs/robots-in-the-oil-and-gas-industry

https://www.accessengineeringlibrary.com/content/book/9781260135282/back-matter/appendix7

Chapter 5

Downstream sector with machine learning

5.1 INTRODUCTION

The downstream sector involves refining the crude oil obtained and converting the raw material to the final product and supplying it to customers. The major portion we can see is the refining and the processing of crude oil and natural gas to the value-added products of our day-to-day life. The marketing and distribution sections are covered by the downstream section. The products such as gasoline, diesel, liquefied petroleum gas, asphalt, and other petrochemicals are valuable chemicals and the raw material for various products. Ethylene acts as the raw material for polyethylene, and methane is the feedstock for methyl chloride, methylene chloride, acetaldehyde, and many others. Any operation beyond the post-production of the oil and natural gas operations is encompassed under the downstream operation and the utilities that are doing that come under the downstream sector. Companies include the refineries, oil and natural gas distributors, petrochemical plants, and the retailers involved. Cost optimization is mainly concerned with processes and unit operations. The main cost is for the energy and the raw crude availability. There is a need to change the process operation parameters differing from crude to crude. The comparatively low valued products are produced when the simple distillation is applied over the denser or heavier crude having the lower American Petroleum Institute (API) gravity values (15–25). While the heavy crude oil having more API gravity values the process becomes expensive to make them more valuable and routinely useful value-added products. Some crude has higher sulfur content which is not desirable in the operation so preprocessing is required and it adds the additional processing. With the help of machine learning (ML) we just not only have to watch and monitor the assets but we need to create novel technologies with the catalysts, and the process parameters so that better accuracy and effective production can take place.

Figure 5.1 mentions the quantitative measurements of the U.S. reiners and blenders petroleum production of 2019.

There is a need for making the enterprise-wide optimization for the refining and the supply chain comprising the distribution and the manufacturing via an emphasis on the incorporation of the various decision-making systems (Shah et al., 2011).

We also goal to perceive the influence of the competitive advantage on profit. The returns on the investment of the equity and the data analysis of the past years are necessary for creating predictable values for the ongoing and upcoming years (Kumar et al., 2017). Furthermore, we need to work on the smart refining technologies based on artificial

DOI: 10.1201/9781003279532-5

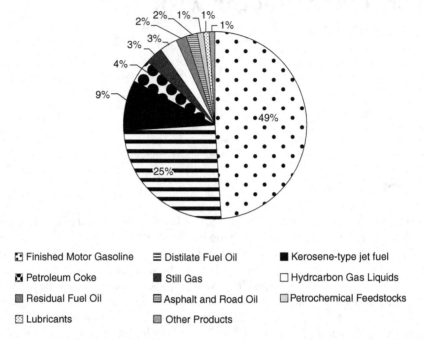

Finished Motor Gasoline   Distilate Fuel Oil   Kerosene-type jet fuel

Petroleum Coke   Still Gas   Hydrcarbon Gas Liquids

Residual Fuel Oil   Asphalt and Road Oil   Petrochemical Feedstocks

Lubricants   Other Products

Figure 5.1 U.S. Refiner and Blender's net production of petroleum products of 2019, 7.46 billion barrels, is encompassed.

intelligence (AI) and ML platforms ruined by the data and the smart algorithms and the software for better performances, operations, equipment, and process designing.

Advanced plant modelling and simulation with process and accurate condition and efficiency, risk analysis, remote operations, conditioning and dragonizing of equipment with machine learning, digital and Internet of Things (IoT)-based services, machine vision for safety, energy, and asset management.

5.2 SMART REFINING PROCESS INTEGRATED WITH THE ML

Smart fabricating and creation processes will transmute the petroleum and petro-chemical sector to linked interconnected, information-driven, and automatized pro-cedures and a complete ecosystem digitally. Utilizing the real-time and the future's cutting-edge technologies with a great valued support system, smart manufacturing empowers harmonized and enactment-oriented enterprise that reacts swiftly to the demand of the market and the usage of the society with the optimized energy and the material supply demands and their usage overall shooting up the sustainable-technological-productive ecosystem with the innovation and the value-added products benefiting the entire supply chain with the monetary basis. The same was the scenario when the world was changing on computer-based working things. ML will provide the platform for all the permutations and the combinations possible with the technology and the applied novel innovations and the development.

Yuan et al. (2017) identified and presented several examples of smart manufacturing for the petrochemical industries like the blunder detection of catalyst and the catalytic cracking unit driven by the IoT and the big data-based ML systems for the advanced optimization and the challenges that need to be conquered for establishing the smart technology. They also identified the opportunities and the challenges to that such as operational agility, the fast response to the new sudden arrived situations for the reason being of the variation of the feedstock, price, and the market demand. Such fluctuations affect the plant's performance Things must be terminated using operational procedures such as process flowsheet reconstruction and pressure, temperature, and flow rate variation.

We just need to calibrate the data in such a format that the exact operating parameters are matched and those particular optimized operation conditions are adapted in just a few seconds. So that there is less time to do the particular task (Chachuat et al., 2009). Abnormal situation management, planning, and scheduling of an entire oil refinery and the adjustable data-driven models are the key future challenges to be conquered, developed, and digitized through ML (Figure 5.2).

Steurtewagen and Poel (2020) in their research proposed novel data-centric options to optimize a catalytic cracking unit for that, they designed a soft data-oriented fluid catalytic cracking unit. They utilized the data from existing sources of the refinery and other similar systems. They used the new sensor (soft sensor) model to track the catalyst in the fluid catalytic cracking unit. They concluded that the model can give higher yield, lesser consumption of the catalyst, and the efficiency of the process is increased. They also claimed that the system can be optimized and the top of the pipeline as smart process control. The aim was to conduct the ML bases saturation level of the catalyst. This leads to a stepping stone for changing and implementing the technology for the manual measuring frequency. The cost reduction was observed in

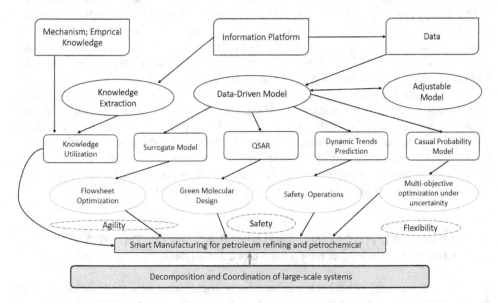

Figure 5.2 Key features of smart manufacturing for the oil refining and petrochemical industries (Yuan et al., 2017).

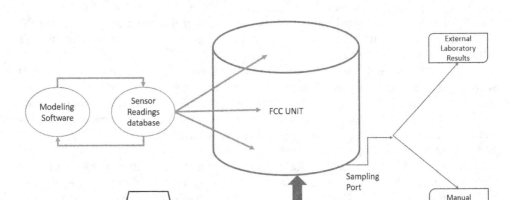

Figure 5.3 Schematic of experiment setup (Steurtewagen & Poel, 2020).

it and it was able to do it. We can further develop the soft sensor data-based system also measuring the amount and the quality as per the crude and giving the results with the report for the particular task. Also, the operation can be handled by the data instructed robots and the digitalized system can be visualized (Figure 5.3).

Min et al. (2019) in their paper presented the digital twin framework for the petrochemical production control and optimization bases on the IoT and ML. They provided the approaches to time series data processing concerns in digital twin modeling like the unification of the frequency, time delay issues, and the immediacy demand. This particular technology-enabled CPU is in the petrochemical industries. The main emphasis was that if the historical data is not able to meet requirements the new digital twin framework can set the algorithms with AI and predict the results.

Li et al. (2015) presented an overview of the smart factory needed in the petroleum and the petrochemical factories. Ubiquitous factory, factory of things, smart process manufacturing, and the Industry 4.0 areas were the key essential work sources needed for the development of the leading-edge smart solutions for the ongoing and as well as the upcoming energy technology. Modeling and simulations, intelligence control, and intelligent optimization algorithms are the important aspects need to be incorporated with the ML tools so that we can have a productive environment.

5.3   ADVANCED MODELING AND SIMULATION OF THE PLANT AND PROCESS FOR BETTER FUNCTIONING

The refinery business is becoming increasingly complex as new feedstocks and market demands emerge with varying properties and newly demand of the industry. The lessening profit margins are also observed in the refineries at the global level. For sustaining the lucrativeness in the business and the operations, the refineries need to focus on process simulations to achieve the best operational excellence. Process simulation technologies have the greater ability to assist and support the operation majorly spanning operational troubleshooting, selection of the crude, planning of the refinery, marginal analysis of the profits, overall paneling, and many more. Advanced modeling and the simulations of the current process helps in the additional entities requirement for the effective operations and intensifications as we can't stop the ongoing running plant and have the practical analysis of the scenario changing the operating conditions which will change the quantity and quality of the product (Mohan et al., n.d.).

Retrofitting of the refinery hydrogen network is one of the important aspects of modern refineries as the stream of off-gases consists of hydrogen and other lower hydrocarbons. Recovering them we can get pure hydrogen and useful lower hydrocarbons. Deng et al. (2020) proposed a systematic procedure for retrofitting the network of the hydrogen and the inter-related processes via analysis of the technical aspects with the process modeling and the advanced simulation techniques. They utilized the Aspen Hysys for it. The total benefits of the light hydrocarbon recovery can hit the target of 8 million CNY/year (Chinese Yuan Renminbi). The remarkable benefits of the monetary gains are seen and also the investment payback period is around 8–10 months from the hydrogen and the Low Hydro Carbon (LHC) sales. Thus, we can make the similar and the other recovery process stimulations in the existing refinery not having such technology and can transform the aura of the profitability and the effectivity of the operations. Simulations not only provide the new designs or the new establishments. Existing units can be stimulated and if troubleshooting is required we can make alternative routes and carry on the operation without shutting the plant. Monitoring of assets and equipment types, as well as their maintenance and operation, is possible with the built on the ML-based models with the high-end expert software available like ASPENTech series and others.

We cannot create every technology and core from scratch; instead, we must adapt it to the best existing solutions available and try to make it out a new approach and the models for the great milestones.

Robertson (2014) in his thesis presented the novel modeling techniques for the simulation, optimization, and examination of the chemical engineering tasks within the petrochemical and the refineries unit operations.

Control volumes, equations of the [material balances, kinetics, energy balances Material Balances, Kinetics, Energy Balances.] initiate industrial process modeling. Thermodynamic relationship with the logistical equations. Improving convergence, consumption saving techniques, and loading and unloading of the crude oil models were the further topics. Optimizing techniques, stochastic optimization, Variable Neighborhood Search/Threshold Accepting (VNS/TA) metaheuristic repairing strategies, stochastic optimization of the distillation column using these repairing techniques, and the operational optimization integrated via production scheduling were the key aspects. The end-last section was the data-driven modeling for the monitoring of the processes. The vital sub-domains were the data-driven approaches for process monitoring, unit application of the polymerization, and the Tennessee Eastman process application. They provided the improvements with increasing computational ability to perform the tasks and exclusive search results for the solution space. Incorporating the data-driven smart models with the ML techniques will give us the leading-edge solutions with the cutting-edge future's technology which can be applied via even small scale to the huge complex industries with an excellent advantage of the cost-effectivity. This technology is new and not in practice, so the cost is higher, but when more work in terms of the research, development, and implementation is carried out by the authorities, this transformation will take place and the main agenda of Industrial Revolution 4.0 will be achieved (ConocoPhillips, 2020).

Zhou et al. (2012) suggested a bootstrap aggregated model tactic for the estimation of the quality of the end products of the refineries with the erratic crude oil. The varying quality of the crude changes the operation process variable and also the operation timings, which ultimately results in the different end products. Soft sensors used are facing difficulties in it. A bootstrap-amassed model based on the Partial Least Squares or Projection to Latent Structures (PLS) inferential valuation method incorporated with the online classification of the crude oil was developed by them for the drypoint assessment of the kerosene. The base of the ideation was creating a solution for estimation model for each oil and selecting the appropriate model for that via classifier on an online estimation. These models are generally utilized for solving the problems caused by the trivial amount of the training data, which is commonly used in the industries. The results showed that the accuracy of these models and the generated data with the similar type of oil and the online data of the oil were higher than any other. The better robustness was seen in the bootstrap aggregated PLS model. The simulation and the practical applications were also sounding perfect. From here the catch is that we can incorporate the model and design the novel approach system so that with the varying feed and the past data it can predict the properties and also the changes required in the process variables and can back-calculate the product quality and quantity of the plant. This can save a huge time and also can be cross-checked with the experts and applied soon.

Ayafor (2018) presented research work exploring the development of a non-isothermal hydrocracker reactor using a lumping methodology or the approach of continuous kinetics. They observed that using the Aspen plus the division of the petroleum into different fractions the estimation task of the parameter analysis and the computational timings are reduced. The finalized model is giving the demanding results for the analysis and the estimation of the temperature and the concentration. They used the Pseudo Random Binary Sequence (PRBS) sequence system recognition stratagem for evolving a better model for the dynamic system. The model was a linearized state-space matrix that was instigated for the Single Input Single Output (SISO) and Multiple Input Multiple Output (MIMO) control. They also pretended to develop a plant-wide control system in the future. The fruits were the elevation of the profits and the maintenance qualities (Figures 5.4 and 5.5).

5.4 REMOTE SYSTEMS OPERATIONS

Reliable operations are key to a profitable and sustainable business. Remote diagnostic services allow the utilities and the companies to support their customers 24 hours a day and 7 days a week increasing the availability and the reliability in terms of services and the positive approach to their solutions. The remote diagnostics and the integrated services are the combinations of monitoring the complete operations of the oil and gas industry equipment. Siemens is one of the multi-national companies currently monitoring operations in more than 80 countries. In their remote diagnostic centers embedded in the dedicated locations of the line of the product, the data has been processed by more than 1,200 units of oil and gas industry power generation and industrial clients. They are serving the complete trains consisting of gas turbines of both industrial and aero derivatives steam turbines and compressors covering the generator sets and

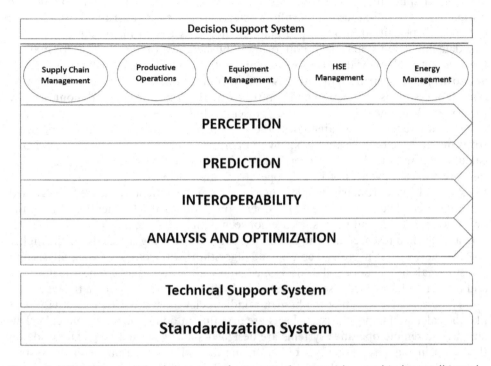

Figure 5.4 The framework of the smart factory in the petrochemical industry (Li et al., 2015).

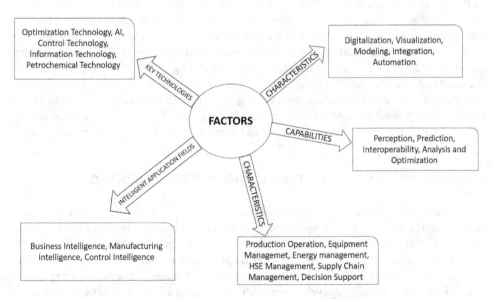

Figure 5.5 Factors of the smart factory in the petrochemical industry (Li et al., 2015).

mechanical applications. The beauty is that the advanced pattern recognition and the normal behavior techniques support fault detection at an early stage and the unplanned outages can be shifted into planned maintenance activities and trips can be avoided. The data collection step is the initial form established on a fixed time basis for making the cycle active then the data is processed and diagnosed. The main task is to save time and to avoid false alarms to the utilities carrying out the operations. The SCADA-based operating system is at function and controlled by the control room and monitored by the experts. With the help of the existing present technology, we can create an integrated platform where the data obtained from the systems can be combined with the operation software, and the one-click away stepwise operation can be planned with the existing calculations and the thoroughly verified past instances of the similar way. We need to improve the effectiveness of remote operations and their working analogy with ML tools and AI Extraordinary complex algorithms and the data network are to be gained and simulated. Then we can create the operations more easily. The development of the novel requires time, and the results are not seen within a year. Also, much such literature and applied research need to be carried out before setting the goals for the higher complex networks. The guided coding and algorithms with complex data analysis and decision-making capacity will be developed, and with the passage of time, the machine and the ML will learn from self-generated and self-run data and perform the tasks.

These tools are assisting the human minds in elevating massive productivity. Just like the calculators and the excel sheets are assisting in doing the mathematical calculations, the remote operative systems are designed smartly so that they can decide on their own in ongoing operations based on the data. The operations with the system and intellective decision-making are important. The algorithms and local connections are needed to establish feasible communication between the server and the equipment. Local-based networks or online cloud communication can also be designed. The internet of factories has the concept of the local external connection for the communication and digitalization of the industries to avoid data and privacy issues. Such a novel network can also be designed for the external within a location for better and optimum and faster communication. In the same network data transfer can also take place autonomously. Indeed, by collecting the manual data, in one shot we can hit many targets. If such a type of local connection or server is available, then it needs to strengthen and make developed and smart technology including the remote controlling and the diagonalizing of the operations in the refinery and the other petrochemical units. Automation has to lead us to develop the existing machinery to run digitally for that the system, and the sensor-based software need to be engineered so that the adoption and the operation of the smart remote-based systems become easier on both sides ensuring safety.

5.5   RISK ANALYSIS WITH THE ML DURING THE REFINING

Risk analysis or assessment is the basic prerequisites of any industry and the petroleum industry is dealing with hydrocarbons which are easily causing hazardous accidents if a minor error occurs. With the increase in the production capacities and the operational enhancements, there are several challenges we need to encounter. The developments in sensor technology, power processing, remotely operational capacities, and IoT technology can unravel remarkable value by eradicating idleness, increasing the uptime, and more promptly allocations of the feedstocks of the raw material. All the operations can work effectively and the economic benefits can be seen when the plant or the refinery is risk-free and the safety of the safe operation is enhanced (Slaughter et al., 2015). We need to develop and incorporate the new cutting-edge technology in the risk assessment modeling and evaluating the system in a sustainable and economic prospect. AI and ML are the two huge domains on which the models and the aggregated system can be assembled, created, or redeveloped on the existing one. The advantages are observed when there are lesser shutdowns, increased production capacity, novel approaches to the processes. There are certain incidents where the initial conditions and the design parameters are different and later on with the time allowance there are certain changes observed in the parameters as the process is in the middle and the winding session. Smart and sensitive risk alarming systems should be assembled to avoid small to huge accidents and to ensure the safety of human life and the protection over the costly assets and massive investments.

Paltrinieri et al. (2019) presented their work on the industrial risk assessment approach based on ML. They proposed a Deep Neural Network (DNN) model for the evaluation of a drive-off scenario in the drilling rig. The development of the model was to predict the risk increasing or decreasing as the conditions of the system fluctuated. Results from the test on the presented model and the comparison with the Multiple Linear Regression (MLR) the previous model show that the former was more accurate for evaluating the dynamic conditions and also possesses the flexible operating conditions to deal with the undesirable as well as the unexpected events. High model sensitivity does not tolerate inaccurate indications and can potentially lead to overfitting. They suggested selecting the models via customization and the prediction over the intended purposes through metrics, tolerances, and the creation. They also added that considering these odds, we can use the DNN model with the safety precautions for the decision-making. Thus, to be successful in running the algorithms and the programs, we need to consider some events and duplicate the odds with the better one. The combinations of such models and the joint operation can give us better results with the assured results. Sensor technology, IoT, and the internet of the factory/plant are important and need to be incorporated. Risk analysis is a huge task and it requires the coverage and the functioning of operations causing the huge disasters. The optimum design of the equipment is usually made for that particular operation. But the lag is in the reaction time. When the sensor responds and the alarm reveals, the human resource has to acquire the particular shutoff or shutdown. Here we need to boost the system in such a way that before the alarm gets ringing or the signals are achieved the model can predict it before and get back to you with the specific things that need to be followed so that a fast decision can take place and the minimum loss can happen. Even though as discussed in the previous chapters the remote operation is required for such operation and the steadfast timings. The human in charge is a few clicks away from the desired conditions. Thus AI works and ML can calculate, communicate, and simulate the desired results.

Ishola (2017) in his thesis presented the interruption risk level of an oil and gas industry and its operational analysis. They developed case studies to consolidate with the results in the thesis to evaluate the risk level of the Petroleum Refinery Process Units (PRPU) operations is applicable to realistic conditions.

They utilized the Fuzzy Evidential Reasoning (FER) methodology for the assessment of the risk levels of the PRPU operation and the adoptions for the quantitative risk analysis for the refinery industry to support audit and for the risk analysis before management of change in the operations. They stated that the application of the Fuzzy Bayesian Network (FBN) technique in the presented work can reveal the convergent effect of the risk elements and the other creations can lead to the disruption of the petroleum industry. The assessment leads to the boosting the FBN technique with a quantitative approach for supporting the risk decreasing things. The Bayesian Network (BN) models are suggested to be utilized to predict the state of the operation of any process unit in a refinery in the way to give the relevant information to support the safer operations. The overall proposed framework developed in the work uses consolidate current procedures for risk management. The FBN is practically working with hazard analysis like Fault Tree Analysis (FTA) to predict the possibilities of failure in the process. There are also some other models which are can be integrated.

5.6   CONNECTING IoT AND OTHER PARTS OF THE DIGITAL DEVICES

The major aim of the IoT technology in the petroleum industry is to connect the physical objects and the internet and to establish a faster connection and communication between the devices and human interference for enhanced operations and accuracy. Wearable devices, vehicles, equipment, and the rooms embedded with the electronic media in the form of software, sensors, and the other electronic media transfer the data to the control room. The ease of transferring and the collection of data without human interactions prove to be a faster and reliable asset of the organization. This data is transferred to the central platform for further processing and operation. The new business model is combining the forward-thinking of the oil and gas organization focusing their IoT technology and converting them to the smart operating systems and the ecosystem focusing on better data management and the ML and AI as the building blocks for the novel creations. There are generally sensors or communication systems established in the refineries for better operation. IoT helps in connecting them all in one under a common ground for the ease of access of the on-time or real-time information and the real-time scenario. At the same time, it will give access to the engineers and the authorities to send the signal and program the operations with better effectiveness. The same technology assembled can be used or a new one can also be established with the economic aspects and the capacity of the utility for the assets. The smart communication system can be put forward with secured communication networks. The International Data Corporation predicts that the worldwide technology spent on IoT technology would be US$1.2 trillion in 2022 (Framingham et al., 2019). Ericsson forecasts that the number of cellular devices connected or used for IoT technology would be 3.5–3.6 billion, whereas the wide-area IoT diplomas would be 4.1–4.2 billion in numbers (Ericcson, 2018). Almost every multi-national company is adopting the IoT technology in which the United States of America is leading, followed by the United Kingdom, European countries, China, and so on. The supply chain companies like Siemens, Cisco, and ABB provide the services and the industrial equipment and the manufacturers are the early adopters of it (Rossmann et al., 2018). Using the IoT technology, the refineries can plan their shutoffs, enhance the safety procedures and the operations, and minimize their downtime. The downstream industry is an ecosystem-operated domain creating the value chains from the economics to the novel consumers by expanding the visibility into the entire supply chain from the feedstocks to the end products. The data will drift in real-time conditions and seamlessly from the ground levels to the energy suppliers. Better decision-making, process optimization, and a safer working environment. Even with the small contribution of AI and IoT, there is a massive growth observed.

(Raynor et al., 2015). Generally, oil and gas companies do not share the information relevant to the operations and the refineries so the global information exchange is creating the speed breakers in the whole overview and the planning. Certain government-owned companies are encompassed have the access to internal communication so that they can practice this and have a prediction for the private company for the same and a nationwide estimation or the prediction can be performed. But the interior ecosystems can be assembled from the unit-wise operation of the IoT-based central control room.

ML-based programmes can be used further for the specific equipment when the steam temperature and pressure exceed the safety levels, so that before reaching the near-safe operating temperature, it can alarm and adequate human time is available for the specific actions. Such as if in the catalytic reforming or the FCC unit the product quality is just below the accepted, thus it can calculate the necessary changes required in the parameters for the proper output it can suggest and implement. Thus interconnecting the info and processing properly with accurate results and on-time operation will serve as the best solution required economically.

5.7 MACHINE VISION FOR SAFETY

Machine vision is the technology that provides the facility to visualize the industrial equipment or any assets which are just functioning on the data, but it can now sense and process data in the form of the image, pixels, and the visual movements of the particles. The key domains observed are quality assurance, robot/machine guidance, calibrations, and testing methodology, real-time process control, collection of the data, sorting and counting calculations, and the monitoring of the equipment. If we see the overview of the petroleum industry and specifically the downstream sector, machine vision is an expensive technology as the good quality of the cameras and the visual technologies need to be established. If we see the main unit, that is a refining unit or the refinery complex. Machine vision can be implied to the end product quality assessment. The technology can thoroughly check and analyze gasoline, diesel, and many others. Furthermore, if we build the algorithms in such a way it can predict the octane and cetane number, then it is far better. Furthermore, it can incorporate the robots and the robot guiding. In the new era of the Industrial Revolution 4.0, we can see that more use of robotics and automation is going to be there. For guiding and monitoring over controlling these robots machine vision can be employed. As the ecosystem of the AI-based connected system is going to be designed and assembled the technology will also contribute its expertise. Smart detection over the sensor with accuracy will be the key benefit. Robots with smart vision can understand the shapes, volume calculations, and packaging or the delivery aspects. ML will be the main programming and the development of such interconnected domains with smart devices and such costly technology. Also, cyber security and the safety of such equipment are going to be the concern Safer algorithms and a good amount of a safe environment are developed while designing or creating the programmes so that the company's treasure in the data format is not misused.

A collaborative R&D with the manufacturing unit and the technology provider efforts must be made for the mutual benefits and converting the concepts to reality. Also, there can be an option where the scrap or the useless cameras can be taken and made customized programming rather than getting the new ones. There can be such *n* number of permutations and combinations which can be performed and done on a trial and experiment basis. A manufacturing unit or the refinery can have the in-house development of such technology for better and safer purposes. Also if the company has the first forward integration business model like the shell has its own refinery and also they have the outlets of the petrol or diesel for the end users. They can establish machine vision technology at the pump. Where the digital data collection and the payment and the quantity assessment can be performed so that the capacity and the trustworthiness of the customer can be known. Also, the supply chain management companies can equip or develop for managing the local transportation services and the logistics for the products. Also a business-to-business model can have it for the inter-demand and the supply of the intermediate products for having the better logistics which can connect to their central server and have the information for the real-time scenario and the future perspectives.

5.8   ENERGY AND ASSET MANAGEMENT

Energy is one of the key ingredients or the basic premises required by any industry. The oil and gas industry is one of the producers of the energy or the fuel we can say. However, energy is used in the creation and chemical operations to fine-tune the product's energy. The asset is the investments done by the company in the plants for better functioning. It can use expensive machinery, software, smart technology, human resources, and each domain in which the initial investment has been done to benefit the company. Energy and asset management are the two huge domains. Like the high-temperature sensitive reactions in the chemical processes, a little change can cause more difference in the results. In the same way, a modest change or the proper management will affect the overall consumption of energy, and the life for the assets will increase.

We discovered that carrying more things at once is the refining complexes at the downstream sector-main junction. Refineries generally use electrical energy in the process. Most of the process equipment and the process are consuming the power directly or indirectly. Proper management is very important. About 60% of the operating cost are energy costs. Energy efficiency is a fundamental part of the refining and petrochemical industry. Asset management in the energy sector is a complicated process, and the results or the outcomes are the better quality and higher reliability of the services provided and the better optimization of the infrastructure maintenance costs. The new production management systems are designed on the principles of the equipment maintenance-based condition, the failure probability of each distribution of the energy. Risk-accountable asset management system covers the aspects of the economics, environmental and reputational risks.

The phrase "what if" has a lot of power: what if you can predict problems before they happen? What if you could cut down on unplanned downtime? What if every machine and piece of equipment is operating at peak efficiency while remaining safe and secure?

For getting started one needs to be connected to capture and manage the data to gain a more complete and accurate view of the equipment. The first step we consider is monitoring the machine and equipment health. Next is to use advanced analytics to get the insights to predict and diagnose issues early and then respond to them before the failure. This insight helps in making better decisions. This domain is reliability management, and the part of maintenance optimization is to optimize maintenance strategies that balance reliability performance and costs.

The catch is to provide the right information in the right hands and at the right time. There is the need of designing and creating a unique model for both the domains covering asset and energy management. With the help of ML, AI, IoT, cloud computing, and the sub-aspects of Industry 4.0, we can make a smart system covering the domains accurately and effectively. The system should be flexible and smart enough to adapt to the global changes as well as counter the local ones too. It should be retrofitted in the running units and can give the desired results. Assets require maintenance on a time-to-time basis.

Human error is inevitable, but when it comes to a crucial piping process, it becomes more important for the detailed analysis. It's part of asset management to look for the complex network of the pipes and properly plan their regular checkups and the scheduling maintenance on time. There are also some operations in which energy intensifications are required, and if the proper dynamics control is performed, greater energy can be conserved and we can make the savings.

Sayda and Taylor (2008) demonstrated a better progression of the development and the designing of the intelligent control and asset management system. Their system prototype design was prolonged combining more agents and functionalities to cope with the complex scenario at the huge complex systems in the industry. They also suggested the system deployment scheme to conduct the real-time system simulations. They believed that the proposed system will be much more useful in the existing utilities.

Furthermore, for the same behavior, we can use a modified version of the artificial intelligent pathway. And make a comment about it in the actual scene.

5.9   REMOTE OPERATION AND PERFORMANCE SHUTDOWN USING IoT

The era of mobility and connectivity has enhanced user-focused designs in each aspect of the system. There is an imperative need for the concept to be applied to industries, with a focus on the traditional and dominant energy industry, i.e., oil and gas industry. Presently the global refineries are highly instrumented and processed in real-time on the scale of milliseconds. Programmable logic controllers and distributed control systems have evolved in a significant manner to survive up with the ever cumulative demands of the operations and the operational demands.

Shutdowns are temporary, which means that it has an explicit start and finish while the operations are ongoing. Operations involve work that is continuous without the specified time. There is a preferred and planned execution of the shutdowns for the refinery for the maintenance and the assessment of all the machinery and the equipment. Shutdowns or turnarounds typically cost about 30% of the annual maintenance budget, and any delays will result in additional costs and a delay in that will cause the cost for it. Shutdowns are dangerous, the utmost care is required while performing it, and the same care is required in the commencement (Sayda & Taylor, 2008).

Using the IoT and AI, we can create such an advanced model and the ecosystem such that the preplanned detailed planning and scheduling is done by the system. With the big data and the advanced data analytic tool, automated planning can be performance-wise shutdown planning based on the past data available and the optimum values from the theoretical calculations and the practical purposes. Remote operations are generally performed and the enhanced versions and the pathways are mentioned in the above Chapter 4. The main goal is to design the systems and make them retrofitted into the industry without lagging and disturbing the production or ongoing operations. Shutdowns are performed under a set of conditions and the parameters as per the standards at the National and local levels so they can also be incorporated into it. Also, our focus is to carry this mission in the economic aspects without spending much money and creating the wealth for what we have that is the key ingredient for it. For creating such novel ingredients we need to work in the industry and have the access to some requests for fulfilling it. The critical data is not being shared by the companies. For that, the company had to take the initiative to carry out things by themselves and consult the consultancy companies. Data privacy is an important parameter that needs to be taken care of at each moment. As data is going to be the new currency for the future. Research and development teams are focusing and doing but sometimes they require help from the software and the IT companies. Collaborations should always be welcomed, and it is practiced and enhanced with more enthusiasm and the passion for getting success.

REFERENCES

Ayafor, C. (2018). *Dynamic Modeling and Advanced Control of a Refinery Hydrocracker Process Dynamic Modeling and Advanced Control of a Refinery Hydrocracker Process, Graduate Theses. West Virginia, U.S.: West Virginia University*, 1–58.

Chachuat, B., Srinivasan, B., & Bonvin, D. (2009). Adaptation strategies for real-time optimization. *Computers & Chemical Engineering, 33*(10), 1557–1567.

ConocoPhillips. (2020). "ConocoPhillips' board of directors approves spin-off of Phillips 66". https://www.conocophillips.com/news-media/story/conocophillips-board-of-directors-approves-spin-off-of-phillips-66/. Accessed Aug. 4, 2020.

Deng, C., Zhu, M., Liu, J., & Feng, X. (2020). Science direct systematic retrofit method for refinery hydrogen network with light hydrocarbons recovery. *International Journal of Hydrogen Energy*, https://doi.org/10.1016/j.ijhydene.2020.05.029

Ericcson. (2018). Ericcson mobility report. Tech. Rep., Ericsson Technology Review Articles. Swedan: Ericsson Company.

Framingham, M. (2018). *"Idc Forecasts Worldwide Technology Spending On the Internet of Things to Reach $1.2 trillion in 2022,"* Online. Accessed Oct 3, 2019. https://www.business-wire.com/news/home/20180618005142/en/IDC-Forecasts-Worldwide-Technology-Spending-on-the-Internet-of-Things-to-Reach-1.2-Trillion-in-2022

Ishola, A. (2018). Advanced Safety Methodology for Risk Management of Petroleum Refinery Operations. U.K.: Liverpool John Moores University.

Kumar Sar, A. (2017). Competitive advantage and performance : An analysis of Indian downstream oil and gas. *Academy of Accounting and Financial Studies Journal, 21*(2).

Li, D., Jiang, B., & Suo, H. (2015). Overview of smart factory studies in petrochemical industry. In *12th International Symposium on Process Systems Engineering and 25th European*

Symposium on Computer Aided Process Engineering (Vol. 37, Issue June). Elsevier. https://doi. org/10.1016/B978-0-444-63578-5.50009-8

Min, Q., Lu, Y., Liu, Z., Su, C., & Wang, B. (2019). International journal of information management machine learning based digital twin framework for production optimization in petrochemical industry. *International Journal of Information Management*, *49*(May), 502–519. https://doi.org/10.1016/j.ijinfomgt.2019.05.020

Mohan, S. R., Product, S., Specialist, M., & Technology, A. (n.d.). *Five Best Practices for Refineries*. Bedford, Massachusetts, U.S.: Aspen Tech, 1–10.

Paltrinieri, N., Comfort, L., & Reniers, G. (2019). Delft University of Technology Learning about risk machine learning for risk assessment learning about risk: Machine learning for risk assessment. *Safety Science*, *118*, 475–486. https://doi.org/10.1016/j.ssci.2019.06.001

Raynor, M. E., & Cotteleer, M. J. (2015). "The more things change: Value creaton, value capture and the internet of things" *Deloitte Review*, *17*, 54–65.

Robertson, G. M. (2014). *Advanced and Novel Modeling Techniques for Simulation, Optimization and Monitoring Chemical Engineering Tasks with Refinery and Petrochemical Unit Applications*, Doctoral Dissertation. Baton Rouge, Louisiana: Louisiana State University and Agricultural and Mechanical College.

Rossmann, M., Batut, T., Thieullent, A.-L., Brosset, P., Chemin, M., Buvat, J., Kar, K., Khemka, Y., & Hein, A. (2018). "Unlocking the business value of IoT in operations," Capgemini, Tech. Rep.

Sayda, A. F., & Taylor, J. H. (2008). A multi-agent system for integrated control and asset management of petroleum production facilities - Part 1: Prototype design and development. *IEEE International Symposium on Intelligent Control - Proceedings*, 162–168. https://doi.org/10.1109/ISIC.2008.4635950

Shah, N. K., Li, Z., & Ierapetritou, M. G. (2011). Petroleum refining operations: Key issues, advances, and opportunities. *Industrial and Engineering Chemistry Research*, *50*(3), 1161–1170. https://doi.org/10.1021/ie1010004

Slaughter, A., Bean, G., & Mittal, A. (2015). *Connected Barrels: Transforming Oil and Gas Strategies with the Internet of Things*. Deloitte University Press.

Steurtewagen, B., & Van Den Poel, D. (2020). Machine learning refinery sensor data to predict catalyst saturation levels. *134*. https://doi.org/10.1016/j.compchemeng.2020.106722

U.S. Energy Information Administration. (2020). *Petroleum Supply Annual*, August 2020. Washington DC, U.S.: U.S. Energy Information Administration.

Yuan, Z., Qin, W., & Zhao, J. (2017). Smart manufacturing for the oil refining and petrochemical industry. *Engineering*, *3*(2), 179–182. https://doi.org/10.1016/J.ENG.2017.02.012

Zhou, C., Liu, Q., Huang, D., & Zhang, J. (2012). Inferential estimation of kerosene dry point in refineries with varying crudes. *Journal of Process Control*, *22*(6), 1122–1126. https://doi. org/10.1016/j.jprocont.2012.03.011

Chapter 6

Safety and maintenance with AI and ML

6.1 INTRODUCTION

Health, safety, and maintenance are the key elements needed to be taken care of at the utmost cost. A refinery is a place where the vapor is mixed in the air and air pollution is the key element needed and the focused domain. There are more chances where the health is affected majorly. The process dynamics and the design also play a role over here in designing such a process so that the unwanted gases or the flue gases released are in the purified form. Zero discharge is one of the key points in the health domain. Due to the inherent hazards, especially the explosion, fire, and the chemical, oil refineries are firmly controlled workplaces. Plants follow the warnings and safety procedures articulated in the company's Environmental Health and Safety Manual (EHS) manual, which covers all aspects.

Crude oil is a complex mixture of many different hydrocarbons and their chains with the non-constant concentration containing sulfur, nitrogen, and other gases with the salts, trace metals, and moisture content. If a strike of the fire touches, it can cause a huge explosion. The plant and the equipment of refineries are generally modern, and the processes are largely automatic and enclosed (Refineries, n.d.). Routine operations are processes that generally present a low risk of exposure when proper maintenance standards in terms of design, construction, and operations are followed. However, the threat of exposure is always present. As a result of the broad range of complexities and hydrocarbons.

Everything might be needed for the protection of the performance of the maintenance, repair, or installation work in an oil refinery. The workers are required to follow the knowledge available and the training provided to them by designated authorities and the same is required for the engineers and the allotted job. The principal exposures to hazardous substances occur during the shutdowns or the maintenance work since they differ from the routine ones. The need for artificial intelligence (AI) and machine learning (ML) is to better plan and advertise the best facilities required for the proper operation with the proper time-based calculated parameters. AI and Internet of Things (IoT)-based technology will automatically feed and process the data and regular basis will provide the necessary changes and safety precautions. Human errors will be cut down, and data will also be generated which will further enhance the performance of the ecosystem. Safety and maintenance are interrelated things. Plants and equipment will function properly and increase productivity if proper time-based maintenance is performed. With increased productivity and the assurance of a failproof model, safety will be ensured.

DOI: 10.1201/9781003279532-6

Also, safety is the design aspect and also considered in the designing and the manufacturing of the equipment. The plant design operations are also considered. Generally, health issues are also considered, and ML with AI will help in diagnosing the flue gases and the other vapors before being released to any other utility or process. If we see the Reliance Industries Limited and their Refinery complex at the Jamnagar, Gujarat. The grass-root refinery can handle any type of crude from the globe for refining.

The process technology is designed in such a way that it is secure, and the large mango plantations aid in the capture of hot gases. The need for AI and ML in retrofitting them in the industry is critical in terms of safety as well as the conservation of energy process intensifications. The novel technology and the approach will help in converting the technology and elevate economic and sustainable development. Sustainable development and economic development are dominant. Sustainable development will be in the co-domains of the zero discharge of the liquids and the gases with the prevention of the generating of the greenhouse gases which is the cause of global warming. Utilizing the technology generation and the conversion of greenhouse gases and the other waste materials into valuable products is also the aim of the Industrial Revolution 4.0.

6.2 DEEP LEARNING RISK DETECTING AND PREDICTIVE DIAGNOSTICS

Risk detection and predictive diagonalization of frauds in complex pipeline networks in the entire petroleum industry ecosystem are critical. The detection is primarily required due to the risk of hydrocarbon ignition if proper precautions are not taken.

AI using ML tools and unique and novel technology can be created and assembled with accurate and the desired results with lesser economic investments and higher ROIs. Predictive maintenance and diagonalization help the industry to look after the machinery and the equipment. Equipment crash is one of the foremost initiates of unexpected and unwanted events like the discharge of the HC. Technical crashing is the commonly caused unsatisfactory assessment or evaluation, controlling of the assets technical integrity which leads the incompetence to control the degradation and the release of the hazardous gases or liquids from pressure vessels (das Chagas Moura et al., 2015; IoT and AI allow the data collection for a particular allotted application or the assets without the human interactions in the operations. The collected data by IoT devices can be analyzed and examined by the data-driven technologies to build creative models and values in the different scenarios (Yin et al., 2015). ML applications provide platforms and have numerous benefits, including cost reduction of maintenance, repair stop reduction, ML-based fault-reduction, increased spare part life, inventory reduction, enhanced safety operations, and increased production of the verification of the repairing works.

These techniques and applications help in increasing profits and better operations. The development requires a basic understanding and the process dynamics and control. Flow patterns and flow dynamics of the automated generation and processing of sensor data for further optimization and development of the things.

Rachman and Ratnayake (2019) presented the work in the development of the intelligent system with the ML pathway to facilitate the information communication and reuse and lower the human errors and the result variation inherent in the risk-based inspection screening system. They used the data from the RBI assessment conducted by the four offshore and three onshore oil and gas fields and the processing units, comprising over 200 pressure vessels and 3,000 piping lines. They selected the ten ML algorithms and compared them for better performance. The result of the study showed that the ensemble techniques Gradient Boosted Decision Tress (GBDT) Radio Frequency (RF), and AB were sounding better than single classifiers such as Support Vector Machine Algorithm (SVM), Logistic Regression (LR), and K-Nearest Neighbor (k-NN). The selection of the classifier and the technique was based on the conditions and the requirement but each one required some sort of correction and advancements. The presented work can help the researchers and developers to upgrade it and make it viable in the real fields for benefiting. AI is better in terms of accuracy and precision. The aim was the elimination of appraiser-to-appraiser output variation, by removing the influence of the subjective judgment from the evaluation.

Pennel et al. (2018) presented that ML models are used in artificial lift to detect failure and optimize performance. The work was described for the methodologies and the examples for combining ML, IoT data, cloud-based platforms to augment matter expertise for detection of failure suboptimal performance states in the field of artificial lift. The project also demonstrated the feasibility and success of applying ML algorithms to diagnose and predict suboptimal stages of pumps and their failures. There were problems suspected in the lift in some conditions. Thus one can solve the problem and also utilize the model for the same. Thus, not only in this field, but one can also investigate a similar type of situation or problem faced by the industry and attempt to conduct applied research on it in order to obtain a solution based on ML.

6.3 BOOSTING PRODUCTIVITY WITH PREDICTIVE MAINTENANCE

Productivity is boosted when all the workforce including the human resource and the machine assets or the machinery are giving their best performance and consuming the lesser sources of energy. With humans, the industry is managing from the initiation and the practices are going. The trend has reached where the machines are becoming smarter and they are performing human tasks at a very high and optimal level of accuracy and precise manner. The new technologies suitably increase the productivity of the plant or the necessary changes required in the plant site concerning the processes and the equipment. Also, maintenance plays an important role in sustaining the hike in the profits and also the better performance of the industry. The neural network of the ML tools, AI, IoT data, and sensor- and cloud-based platforms will provide the leading-edge technology in each industry operating at higher volumes and corrosive fluids like hydrocarbons. The model or the system can be novelly designed such that the integration is done of all the applications. The predicted life of the equipment with the workload can be calculated in it. According to that, matching the real-time data of the production with the time, it can intellect in itself and provide the predicted timings for the maintenance required (Çinar et al., 2020). Also, the shutdown timings and the detailed results of the analysis can be calculated. The pumps, valves, and other secondary equipment can also be accounted for in it. As for the major, once a data is available in any format, it must be adopted in the model or system designed so that the accuracy and real-time scenario can be matched and the entire life of the equipment can be used

(Custeau et al., 2018). Productivity and maintenance are interrelated terms. If maintenance is done properly, then increased productivity can be achieved. Productivity is the function of maintenance as well as many other things. Newer processes, process integration, process intensification are the advancements that can be practiced on the technical side. But that generally does not disturb the ongoing operations. In the very tough economic time, oil and gas businesses need an urgent action for the operation at the highest level of effectiveness and efficiency, while at the same time one needs increased production capacity and reduced operational as well as the controlling costs. There is a need for limiting downtime, minimizing the risk of losses, and also ensuring the safety and environmental impacts. From the perspective of asset performance management, the companies and utilities are leveraging the industrial data and advanced analytics to keep equipment running in the safer zone and reliable way in the long run. This collection is used for further processing and the application of the predictive maintenance execution, empowering personnel to act before the failure occurs (Ralph, 2015). Algorithms can also be designed, which can give the manufacturing enough time to create and execute them in the running process so that the shutoff does not occur.

Not only in the refinery it can also be retrofitted in the upstream, midstream, and downstream sectors. The companies may have already installed the technology, but the need is to drive at the maximum run giving the maximum profits and consuming the lesser energy and the economic aspects. A smart comprehensive predictive maintenance program should be designed by the companies that are required for the operations in all three sectors. The preventive maintenance strategies assist and result in the maintenance programs and results for the work to be performed on a fixed schedule or based on the statistics of the operations, as well as the manufacturer's recommendation and comments based on failures and suggestions for increasing the life run of the process and the equipment (Rio 2015).

This can be managed in the enterprise asset management or computerized maintenance management system. This individual technology can be integrated into one program, and this program can also be tailored and supplied to the other companies as per the requirement. Thus ML and AI can be very useful tools in the petroleum industry. Stabilization-based systems can have incredible results if they are planned and used wisely. The integration and application are required.

6.4   DIGITAL PRE-FIRE ALARMING SENSOR MODE

A key constraint of the fire aspects of the fire protection system is the identification and the development of the fire emergency alarm system promptly. At present, the systems are designed on a smoke detection basis. This sensor can sense the fire and alert the place. Generally, it is advised to prevent smoking and such types of activities in nearby areas, and the operation and processes and process design have a major role in preventing the hazardous situation. But for safety the system is essential (Wu et al., 2018). The advanced pre-alarming systems can be modified based on ML tools and AI. Based on the estimating incidences, the fire point and spark point, as well as the safety parameters for the specific unit and piping network.

The safety temperature and the pressure values can be assigned in the system and if it reaches the limits it will alarm the control room for the same. The advanced technology can be simplified in the system by making it smart with the use of AI algorithms. Before alarming the system it can make sense of the data, process it, and take the necessary actions for the same in a very less amount of time. The ML tools can make the calculations in micro or nanoseconds and give the results which can help AI for making the decision. IoT systems can help in faster communication and faster transferring of the signal. Also, the sensors can be fixed in the particular equipment and all the systems are in a loop under supervision. The in-line data collection aids in the feed for the ML programs, and based on that regular-based collection, the system can be further improved and extended in terms of accuracy and effectiveness (Park et al., 2019). Generally, these incidents are not happening due to better care taken by the engineers and the companies, but in history, the numbers and the extent are very much higher.

Gong et al. (2019) presented the work in which a fire detection method based on clip flame multi-feature fusion was proposed. The key advantages were the integration of the color detection and the motion detection which improved the velocity detection based on the improved frame difference method. The fire processing stage was combined with the subsequent advanced feature extraction of the flames to decrease the calculation time. The color feature of the flame was used to detect the motion, color, area change, shape change, and newly developed algorithms of flame centralized stabilization-based spontaneous reaction systems. They also utilized the vector machine for better results. Their accuracy was found at around 95% which proved that their proposal was a novel work having higher accuracy and stability. But the limitations in their algorithms were that the dress color code was also having the same color so it was unable to distinguish between them. It looked like the fire was blocked in the equipment. Further advanced optimization is required for the detection with this model. Thus with the installed camera system, we can incorporate this model and optimize it with the advanced algorithms so that it can give better results.

Also, the smoke-based data can sometimes cause false signals, and the model's accuracy can be dropped down. Wireless sensor networks can improvise it and the intelligent smoke alarm system and the sir condition can also be more accurately diagonalized. The industry can incorporate any model which is suitable in terms of retrofitting and economically suitable for the industry. The model or system should be smart enough for the detection and shall give the results in such a way that the running time is achievable and the hazardous event can be stopped primarily. But it can also be smart enough to give the solutions for the prevention of the same and also the further suggestion for manufacturers for preventing the same issue again. Thus ML and AI and smart technology can save millions of properties and lives.

6.5 MOTOR VEHICLE SAFETY AND IN-VEHICLE MONITORING SYSTEM

Generally, these systems will be utilized for monitoring the vehicles used for transportation and reducing the risks of hazardous events. As the crude-containing vehicles used for transporting oil from the field to storage or processing facilities.

Their safer transportation is much more important in terms of economics and the overall safety of the driver and the locality. The sensor-based technologies are established for the monitoring of mouse vehicles by companies like Shell. But the need is to make it more optimized and smart technology based on AI and ML for the safer and detailed monitoring of them. Also the same is used for the transportation of the product to the fuel pumps and other utilities for the endues. This will create the data and the data is used for the ML tools and the system for improving the cost, maintenance of the vehicle, driver's statistics, and the safer journey. Also, the systems can be made smarter in such a way that wearing a seat belt, driving in the given speed ranges, parking and the spotting of vehicles, usage of mobile phones during driving, and such minute details can also be noted which can help in assessing the drivers. If we see from the organization side, it can create value in terms of the overall monitoring of the single drop of crude to gasoline and try to manage the data which can give the future forecasting and the prediction also helps in the improvising the functioning and planning. The setup is needed to be assembled in the vehicle and it is connected to the server room or the monitoring room. If the allotted task is completed the system itself notes the details for the same and synchronizes the data in the system. Also, live GPS monitoring can be done if a huge volume is to be transported. The design and connectivity are required for monitoring, for which the driver's mobile phones can be used or given to them with the installed application providing all the details.

Such developments are required for making the operations smarter and safer operations. The data sets generated can be used for the better optimization of the routes, fuel costs to the vehicle, and many other things. There are chances when the demand rises and the quick or urgent need is been required by the vehicle of the transportation. So the drivers may be followed the on-field instructions (Retzer et al., 2015). So, the system should be smart enough to detect the needs in the advance and arrange the vehicles available nearby and avoid overdriving. Algorithms and ML technology should be faster and more efficient so that they can sense, calculate, and make arrangements more quickly. The initial stages required human interactions for monitoring and assessing the models. But with the time and the data, it should sustain itself and takes the right decision based on the AI and the algorithms. So, the human interactions and the communication times can be lowered. Then also the human monitoring must be preferred for overviewing things. The demand and supply management can also be done using the data generated by the vehicles and further planning can be done for the management.

Companies such as Shell, Chevron, and ExxonMobil are forward integrated, as are those in charge of oilfields, storage, and refining. The designed or proposed systems can be helpful to them in a greater way for enhancing their operations and reducing the gaps between the communication and the data creation. The technology providers can also tailor the technology to the private sectors and give them the technology. Also, the company itself can develop it in-house. This development cannot be completed in a month because it necessitates numerous trials and complex coding. It can be retrofitted in industries other than oil and gas. Such development is required for better optimization from the smaller to the huge phase of the operations. Thus, in-vehicle monitoring systems should be equipped with the utility function and deal with massive volumes of crude from upstream and downstream domains to the end-customer, assisting them in making future decisions.

6.6   CLEAR VISION COMMUNICATION AND MONITORING MODEL ON SITES AND EXPLORING SITES

The world is moving to digitalization and data is the new currency in today's and tomorrow's world. The communication and monitoring of the exploring sites are requisites needed for the overview of it. Drillers and the utilities must know the exact locations and the clear depth insights for exploring the crude. The setup and the systems are needed to be transferable to the oilfields. The setup is generally built at the base from where the exploration is done for connectivity and communication. The main control room of the company is in contact with the base and provides the required information and guidelines. The automation of periodic reports, Daily drilling reports are required to be generated and kept on file. ML tools can assist you here.

The auto-filled data generated can be filled in the formatted one and the workload of the human resource can be reduced. This can be a great initiative if the data generation can directly feed into the system and auto-generated daily exploration reports can be forwarded to the higher authorities and the system for the records. Similarly, real-time monitoring can help out in uninterrupted flows. For example, if any obstacle is striking in the flow from the pipes, the detection action system can sense it and give accurate results for the same (Paper, 2017). IoT and the internet of factories can also be integrated with it giving a glimpse of the interconnection and the faster processing of the signal data. The monitoring model design shall also comprise the safety of the field's workforce. The insight and the monitoring are necessary while they are in the onshore fields. If anything goes wrong, the information shall be passed in very little time so that the proper rescue operations can be carried out and the lives can be saved with it. The smart system can be retrofitted in other industries, too. The base algorithms may remain the same but tailoring is required as per the conditions and the field of the operation. There can be systems that are discussed here that may be in existence and the work field but the need is to transform them into a self-operating and self-decision-making system based on the power of AI. Pumps are the mainly used devices on the sites where their monitoring is almost important as their stoppage can reduce the rate of the operation. As the sensor-based technology is suggested in the previous chapters the real-time data analysis for the pumps is necessary for the all-time monitoring model to be established. Also, onshore monitoring, environmental impacts, and monitoring are required. To ensure the leakage and other harmful impacts caused to the oceans. Generally, the test is performed by testing the quality of the water and ensuring that the environment is not harmed (Paper, 2020). With the historic data available and knowing the information for the same, the analysis becomes faster and we can make it more by the ML algorithms and the AI. Also, the testing methodology can be changed and make it better with the advanced technology giving better results.

.Offshore exploration sites, sucker pumps can be digitalized and connected to an IoT-based one platform where manual work can be reduced and auto-generated data can be formed, and then the assigned work can be completed and preplanning of the operation can be performed based on how things are going.

The need for developing smart systems is the making of precise decisions on time to achieve greater efficiency and effectiveness in the work. ML can make things easier for us. However, the requirement is to create the items as well as the development of the application and their inter-integration. Converting existing systems to smart systems is also an option.

6.7   SMART HELMETS AND OTHER SAFETY EQUIPMENT FOR THE WORKERS

Workers' safety in the fields is the most crucial parameter for the companies working especially in the upstream domains of the oil and gas industry. There is no doubt that the oil and gas industry is a dangerous workplace. There are numerous incidents that occur, but they are unable to be brought to the public's attention with the correct numerical values. The life of the person is more valuable than the money. Also in the downstream sector, safety equipment is used. Not providing the proper PPE equipment by the company can also be fined for conducting this kind of behavior. But the oil and gas industry is responsible and caring as the majority of them are providing safety equipment. The need for smart accessories is that the primary goal of the PPE of providing safety is achieved. The instruction can be provided if the calamity of the situation is out of drastic. Knowing the location using the smart devices the rescue can be carried out. Also, the real-time data and the monitoring can be done if proper accessories are attached to them and connected to the server (Shu et al., n.d.).

Smart helmets have arrived in the market and are used by many companies on exploring sites. They are equipped with cameras, sensors, connective devices, lights, and other required sensors as per the task-oriented. The risk-based smart sensing devices should be incorporated so that any risk coming to the sensor can detect and give the signal to the worker in the form of audio and the others. We can also use the AI over here and make the signal to their choice and the regional language and instruct the necessary steps to be followed. Other safety equipment is protective communications, disposable respirators, fall protection, head and faces protection, hearing protection, PAPR & supplied air respirators, and protective apparel.

Kim and Baek (2021) presented the study for the development of proximate warning systems that used the smart helmets to receive BLE signals from the Bluetooth device and give the proximity on the visual basis to alert the workers and the other force. They were designed for the mines. The Tx power and the dBm signal were experimented at certain distances. The workers were free and ensured security with the smart helmets. The location and the further things were carried out by the NASA-TLX method. The smart helmets were used and the overall workload was found to increase as compared to the smartphones-based proximate warning system also with the smart glass-based proximate warning system. Further work can be carried out on Arduino-based sensor techniques. Further the sites' operating parameters like humidity, temperature, pressure, and other flue gases concentration can be measured with advanced sensors. We can also utilize it in our oil and gas industry and imply similar algorithms and techniques. AI can give special comments and suggestions to the workers on-site if anything special care or special operation is carried out. As mentioned above, the signal can also be converted to the local/regional language. Thus the guided program can also be developed via media of the smart Personal Protective Equipment

(PPE) kits. Thus ML- and AI-based smart technology can change the scenario benefiting the safety and also the data collection for the optimal operations via integration of the available or novel techniques. These helmets and other PPEs can also be utilized in petrochemical plants and also hazardous events.

6.8   SAFETY WITH THE SECURITY OF THE DATA WITH ML MODELS

As in the previous chapters, more emphasis is given to the data. Data is one of the most important and expensive assets created by the company over the years of operations. If we evaluate its value, it's equivalent to the economy of the plant. Not only in any specific industry but over data security is one of the major concerns that need to be the point of attention or concern over the globe. Its security is critical for a company that is using a new technology that is based on data. Data can also be leaked or practiced in an unethical way. So, firms are concerned with implying the models, and it can also be the reason that many companies with the fear of security are having some resistance to accepting the data-based models mainly ML and AI, which are mainly dependent on them. The developed systems, models, or tools can also be hacked and practiced unprofessionally harming the growth and the images of the firm. Therefore the immanent and valuable work is required in developing the security models in each industry. In the previous chapter, we have mentioned that the firms are needed to take the initiatives for developing the tools and the systems in-house rather than outsourcing from the technology suppliers lowering the risk of the security at the decreased levels. Sometimes due to many complex algorithms and interconnected models, if a single thing is wrong, the entire plant or the firm can face hazardous and undesirable incidents. Also due to huge complexity, it's harder to provide security to each one. But for that such strong and trustable security patches and the software are required to provide the utmost security to the data as well as the cloud-based systems which are functioning on IoT- and cloud-based platforms. Constant research and development are going on in this field for creating such models or the software providing security to the assets.

Defensive techniques are required as the solution for the protection against such issues securing them from hackers and unprofessional conduct (Rio 2016).

New security threats are constantly emerging and we need to look after both ones. Although a large number of frameworks, algorithms, and optimization mechanisms have been proposed by researchers and developers for creating such interesting things.

To conquer such issues a strong defense system or methodology is required from the overall as well as the data security point of view. As with ML, toots are more based on the data quantity and the quality strong foundations are required for covering up the scenario. Currently, the formally standardizing security evaluation methodologies on ML are still at the initial stages and need to be solved and made into practice. Secure deep learning is a novel growth point in the security field. The current works showed that the counterintuitive characteristics of DNNs affect their security. Here is the tool for developing anti-hacking systems with ML and AI tools. If there is any disturbance found in the system, the system can detect it and block it there. The ML and AI models can be used as a helpful tool for the same as we can develop a deep and strong defensive system for the security applying them and also protecting their assets running from them. The unit-hacking or we can say defensive systems are complex in the designing but as the model based on them is the same we can simultaneously develop the security patch models in line with the systems proposed so that the security is created. Also, the entire package like the antivirus software can be developed for the particular systems, overall computer-based system, special security to the data available, and the cloud security for the same. The development is going on, and this work can help the developers as the motivation as well as the building block for the development of the security systems. In the coming years, we all must be observing the theories and the research based on the paper in real-time usage. Thus, ML tools would aid in technology-assisted as well as defence and security models (Xue et al., 2020).

6.9   PIPELINE LEAKAGE MONITORING SENSOR-BASED SYSTEM

The United States faces a loss of around $10 billion per annum for the leakages and the wastage of the oil. The utilities functioning in the upstream and the midstream of the oil and gas industry forecast noteworthy benefits in improving pipeline safety and dependability. The statistics showed that around 94 leaks per mile are observed around 2.5 million miles in the United States. The world is dependent and banks on the oil and gas industry being one the largest contributors to energy generation around 60% approximately. More human resources and hardware-based assets and utilities have been developed but the promising results have not come. The ML- and AI-based smart technological solutions are interconnected with the sensors and the real-time cloud-based platform monitoring to the end-to-end networks. There are established companies such as Biz Intellia, which provides the one-to-one IoT solutions to the industry and the particular domain. Real-time sensor-based simulated data is provided and instant notifications or alerts are sent to the utilities involved in the operation. The utility can fix the leakage with the due time and the cause is minimized. ML with AI and IoT-interconnected networks can be applied here to back-calculate the distance approximated from the few factual points and cross verify with the real-time scenario. The technology can also be assembled in the onshore piping systems. The water sustaining sensors can be used and programmed so that if the underwater leak is there one can find it, detect and apply the same technology. IoT and the interconnected assets will bring the era of digitalization and automation to the oil and gas industry. Digital leak detection systems are also in the practice of working in the industry monitoring the gas transmission and the detection of the leakages, storage facilities, and many more. This IoT sensor-based leakage detection technology can also be integrated with the properties measuring techniques such as the pressure, flow rates, temperature, and other quantities to know the further pretreatment required. These leakages start from the pin-size hole facture and due to such large flow rates and the viscous fluids, the holes get wider. The smart technology developed shall optimize at such a level that can determine the minute losses. The fast leakage systems can be identified using a video camera equipped with special filters that operate at specific wavelengths.

Hydrocarbons tend to absorb the infrared radiations from the respective wavelengths and the detection can be found. Fiber optics can also be used for monitoring the leakages of the changes occurring on the physical changes of the site.

REFERENCES

Çınar, Z. M., Nuhu, A. A., Zeeshan, Q., & Korhan, O. (2020). *Machine Learning in Predictive Maintenance towards Sustainable Smart Manufacturing in Industry 4.0.* MDPI Journals, Switzerland.

Custeau, K., Management, A., & Summary, E. (2018). *Predictive Maintenance for Improved Performance in Oil and Gas.* Oil and Gas Online, United States.

das Chagas Moura M., Lins ID, Droguett EL, Soares RF, Pascual R. (2015). A multi-objective genetic algorithm for determining efficient risk-based inspection programs. *Reliability Engineering & System Safety, 133,* 253–265.

Gong, F., Li, C., Gong, W., Li, X., Yuan, X., Ma, Y., & Song, T. (2019). *A Real-Time Fire Detection Method from Video with Multifeature Fusion.* Hindawi, United Kingdom.

Kim, Y., & Baek, J. (2021). *Applied Sciences Smart Helmet-Based Personnel Proximity Warning System for Improving Underground Mine Safety.* MDPI Journals, Switzerland.

Paper, V. (2020). *Digitalization Roadmap for Indian Exploration and Production (E & P) Industry.* Ministry of Petroleum and Natural Gas, Government of India, New Delhi.

Paper, W. (2017). *Digital Transformation Initiative Oil and Gas Industry.* World Economic Forum, Geneva.

Park, J. H., Lee, S., Yun, S., Kim, H., & Kim, W. (2019). *Dependable Fire Detection System with Multifunctional Artificial Intelligence Framework.* MDPI Journals, Switzerland.

Pennel, M., Hsiung, J., & Putcha, V. B. (2018). *SPE-190090-MS Detecting Failures and Optimizing Performance in Artificial Lift Using Machine Learning Models Analytical Workflow.* One Petro, SPE, United States.

Rachman, A., & Ratnayake, R. M. C. (2019). PT US CR. *Reliability Engineering and System Safety.* https://doi.org/10.1016/j.ress.2019.02.008

Retzer, K. D., Services, H., Hill, R. D., & Services, H. (2015). Land transportation safety recommended practice guidance note 12. https://doi.org/10.2118/156535-MS

Shu, L., Li, K., Zen, J., & Li, X. (n.d.). *Demonstration Abstract: A Smart Helmet for Network Level Early Warning in Large Scale Petrochemical Plants.* 390–391.

Wu, Q., Cao, J., Zhou, C., Huang, J., Li, Z., Cheng, S., Cheng, J., & Pan, G. (2018). Intelligent smoke alarm system with wireless sensor network using ZigBee. Wireless Communications and Mobile Computing. 2018 1–11.

Xue, M., Yuan, C., & Wu, H. (2020). Machine learning security : Threats, countermeasures, and evaluations. Wireless Communications and Mobile Computing. *8.*

Yin, S., Li, X., Gao, H., Kaynak, O. (2015). Data-based techniques focused on modern industry: An overview. *IEEE Trans. Ind. Electron.* 62, 657–667. https://doi.org/10.1109/TIE.2014.2308133

Chapter 7

Finance with ML and AI

7.1 GENERAL INTRODUCTION

Finance is a major domain for running any industry. All facility development and human resource management. Machines, energy, materials, and so on are all reliant on finance. The gaps and the smooth transaction with the given time duration are some of the prerequisites of any running business. Proper financial management is also one of the key ingredients of any successful business model. In recent years, there has been a swift spurt of disruptive smart technologies such as artificial intelligence (AI) and machine learning (ML) in finance because of enriched and value-added software and hardware developed. The finance sector expressly has been a precipitous upsurge in the usage and the use of the ML application to encroachment better outcomes for both consumers and businesses. From accelerating underwriting processes to model validation, market impact analysis, portfolio composition and optimization, Robo-advising providing alternative credit reporting methods, and much more. Enriched revenues owing to better productivity improved user experience. Reduced operational costs as a result of process automation, armored security, and improved compliance.

The ML tools work on the significant perceptions and acumens from the raw data available and provide accurate and precise results. The given information and data available are then used to solve complex and data-rich problems that are critical to the banking and the finance industry. The major tools are the algorithms equipped to learn from the data, processes, and techniques used to find different perspectives and results. Some subdomains are the financial monitoring, risk management programs based on AI and ML, decision-making using the smart and logic-based prediction, making the investment predictions for the funds available, algorithm trading, customer service level improvements with ML, financial advisory, process automation, customer retention models, customer data and development management, secured transaction, and the available predictions. The base ideation is to predict and make the payments of the bills and the dues on a timely basis. Using the data available, we can make predictions about the demand and the supply chains and know where the prices are fluctuating. Using the finance models in the specific oil and gas industry, the areas of the applications discussed in the subsections are the factors affecting the finance or the economic costs to the industry, and in the end, product processing and its optimization. The major cost in the petroleum industry is the extraction and refining processes. Energy cost in the refinancing is the major one. It discusses the cost monitoring and adjustments. Using the AI model to forecast growth, trends, and the market.

DOI: 10.1201/9781003279532-7

Price fluctuations of the gasoline, diesel, and the commercially available fuel depends on many factors monitoring and predicting the trends with AI models helps in achieving the basis for deciding the pricing and also the demand and supply management. Price prediction models are built using data and ML tools. Predictions made based on the trends can help in making the price and fixing them. In some major countries, the fuel prices are fixed by the government authorities and these models can be helped by them in the price fixation. For improved performance, financial modelling with data from the past and present is used. Digitalization of the distribution channel and end-customer products, as well as their revenue generation and profit margins Smart supply chain management and economic monitoring Supply and demand management, as well as computational models, management, and decision-making Using data science and machine learning techniques to manage virtual agents and suppliers. Planning, managing the economy of the industry or the plant, and ML can help and smooth the transition to improve operations and financial flow.

The prediction and the forecasting with the real-time data help in analyzing and making the correct financial decisions. AI can analyze the market trends and the prices of fuels. Thus there are lots of opportunities for the developers, engineers, and the industry for creating such models and technology.

7.2   FACTORS AFFECTING THE FINANCE COST OF THE INDUSTRY, END PRODUCT, AND PROCESSING

Parameters influencing the oil and gas industry are divided into sections, such as in the upstream domain, where exploration and production are the two main pillars that require the most financial investment.

As the cost of the exploration and the production is higher than the cost of fuel, gasoline and diesel are also higher. Thus it directly relates to the end-user connection. Thus more the expenditure is done on exploration more will be the cost to the company and the more the end-product cost. Storage facilities are also the major section in which more financial investment is required for developing the facilities and their maintenance All of the crux and data available in the scenario can be grouped and analyzed, while predicting the volumes available and managing the demands assists the authorities of the nations managing the fuels.

Global oil and gas companies spend a huge amount of their economic share on the computational modeling, simulations, and the geographic conditions of the oil and gas fields per annum (Boz and Arslan 2017). From the economic perspective, it is very important and crucial to get the capacity of the field for getting the oil and get the maximum profits generated and achieve the economic efficiency of oil field development and investment projects. This becomes crucial when the prices of the oil are unstable and the economy of the project area is at a riskier position. The key factor which plays an important role is the project evaluation, choosing the minimum economic conditions, determining the returns on the investments, and the planned gross profits from the project (Kumar, 2017). The decision-making requires clarifications in the values and extent of the figures as well as the dependency on the other factors which are essential to the economic criteria. Geological-physical assessment, technological and ecological factors of the production facilities, and the oil fields are important. All the data or the information available can be put forward to the ML model with AI giving the specific criteria and the acceptable limits. So, the proper and optimum selection of the set of indicators and allowance for the evaluation should be the basis for the development of the managerial and strategic decisions by the company and the investors. The indicators of the project and its economic dependency can be varied as per the changes in some external and internal factors, so the losses or the extra calculations should be kept in the development (Espacios et al., 2018).

AI and ML applications can match all parameters with the specified and the applied conditions and the whole detailed report can be formulated with the specified conditions giving the approximation cost, profits, and the time for the project. The analysis done via human resources required more time. However, if the smart technology is designed to collect data digitally without the need for human intervention and other tasks, the price analysis can be performed by making the system smarter and connecting it with the monetary resources and providing the approximation amount to be spent on the project. The size of the company, growth opportunities, asset structures, volatility, and profitability are some generally leaky factors which must be examined. The behavior of the petroleum companies' uses of the leverages decreases as their expenditure capitals, risks, operating income, and liquidity increase. There are many theories like the Theory of the Finance Hierarchy found to be relevant and directly applicable in the models managing the capital structures and the choice of the application for the oil companies (Franc-d & Magdalena, 2021).

As a result, we propose an ecosystem model for financial decisions based on the factors that play important roles in digital data inputs and allowing AI to take the results. So the higher authorities or the investors can make decisions easily considering all the things. This will save a lot of time and the project implementation can work fast. Also the bifurcated and the time-zone wise and the money needed distribution can be developed.

Thus AI and ML can help in the factors analysis and the end-product assessments and financial analysis for taking the big decisions.

7.3 FORECASTING GROWTH, TRENDS, AND MARKET WITH AI MODELS

AI is very stable and gives outstanding results in forecasting the factors. Forecasting the growth, trends, and the markets of one of the largest contributors to the energy sector, i.e., the oil and gas industry, is tricky as there are lots of data and information available. AI models can help in forecasting the trends. The big tech giants such as Google, IBM, Fungen, Intel, and Microsoft are the key players for that. The global AI in the oil and gas industry was estimated to be around 2,050 million USD in 2019 and it is expected and predicted to be sized around 3,400 million USD. Due to the Covid-19 pandemic and the lockdowns, the industry observed a plunge in the demands as everyone was not able to access the fuels and the other demands. The prices of the fuels decreased around 50% since January 2020 in the United States markets. A record low was observed so the oil is stored and the storage prices have also increased. As a result, proper supply and profit management must be implemented. ML tools can analyze and sort data for us, and AI can make accurate decisions and predictions for the supply chain and value demands. The trends can be observed over the seas and in trade countries. Also, the demand per annum demands can be fulfilled and the profit can be earned wisely without making anyone in debt. Thus sometimes refineries can play a greater role in it. They can also analyze the trends and markets and plan their action plans according to the global perspective. This can be achieved by having the news and the correct information available or gained. Based on the figures and the past figures, the model can analyze and predict the trends. But the news and the figures should be accurate. Because the model can accept input from reputable newspapers, articles, oilfield outlets, and other reliable sources.

One must ensure that the source of information is reliable and trustworthy. Forward integrated companies in the business of E&P, storage, refining, and outlet streams find it easier to analyze data and trends within the company. They can also predict their competitor as per the size and the operation. Government agencies can also operate it as they are managing the prices of the fuels. Due to the Covid-19 crisis, governments are facing greater losses for the developing countries and also for the developed countries. The prices of fuels have been hiked to compensate for the losses. So governments use them for the castigation of the markets (global and local market demands) and fix the prices. Such a model should be developed which can analyze the trends and the market scenario based on the prices, availability of the crude, operating conditions, and the market. Accurate information is the prerequisites for a better prediction. Deep-learning and ML algorithms can make it feasible and faster analysis via AI can make them castigation. Market analysis is another important task that must be completed. In order to do so, the necessary knowledge and information should be available in data and in real-time.

AI manufacturers, developers, and technology providers, research institutes, government and government-aided agencies, institutional investors, national and state government organizations, and companies operating in each field, i.e., upstream, midstream, and downstream sectors, will reap the benefits and be the utilizing agents. Thus, using AI and ML models to forecast growth, markets, and trends can aid in predicting the curvature of the future time in terms of monetary factors and their influence.

The major financial decisions need to be taken into account so that castigation can help the investors or the company itself so that decisions in the favor of the company should be taken and it should benefit the firm with the profits.

7.4 PRICE PREDICTION MODELS BASED ON DATA ANALYSIS AND ML

Prices of the crude oil and the end products such as petrol and diesel are crucial factors of any nation's economy as well as their market (stock market). Energy and its pricing are the backbones of the economy of the nation. And oil plays a greater role in the energy sector across the globe. The main factors affecting the volatility of crude oil are the demand and supply of the oil, political climates, populations, and the economic aspects. Crude oil price market prediction is renowned for its obscurity and complexities. Because of the high degree of vacillation, unpredictable irregularity events, and complex correlations involved in crude oil movement. Panas et al. (2000) mentioned that the crude oil prices and the market have strong evidence of the chaos and develop as one of the most volatile markets in the world. Few researchers have conducted crude oil price predictions. The models proposed were the econometric, single AI model, and hybridization. As an early success story, forecasted the daily series of future oil prices using a nonlinear Artificial Neural Network (ANN) model that outperformed traditional econometric models. Because they were built on the idea of extending in multiple directions. Yu et al. (2008) were the first to apply the ML model, employing empirical mode decomposition (EMD) based on the neural network ensemble learning (NNEL) paradigm. Ding (2018) extended the approach to include the final ensemble step to predict oil prices. Yu et al. (2017) proposed an ensemble for castigation approach, which combines sparse representation and the feedforward neural network to forecast crude oil prices. The results confirmed the superiority of using ML and AI models for price prediction. Global acceptance of AI for the oil and gas industry can make the models more accountable. Also, the models are using promising forecasting tools for the completed time series data with high volatility and irregularity.

As a result, hybridization based on need, systems, and results. Thus, using both developed and novel algorithms, we can design and develop price prediction modes, as well as optimize existing ones. Global impacts play a vital role as the oil and gas business is globally connected and spread worldwide. The channels and the algorithms can boost the analysis competitiveness and the advanced direct ML technologies can be adopted and assembled directly with further modifications. Using the AI price prediction becomes easier if we incorporate data science tools and the data science algorithms in it. Al-fattah and Aramco (2019) developed the AI model for exhibiting the capabilities to describe the behavior and the dynamics of the price direction of the oil price volatility with 88% accuracy. The mainly influential factors are the US GDP, consumption, production, OPEC square capacity, storage facility, and many more were incorporated. The GANNATS hybrid model can be also used as risk management and the ML-based tool for other digital operations of the oil and gas industry. Thus, we proposed an integrated model for the price prediction incorporating all the factors affecting the prices of the fuels in the sector on fields. Political basics, as well as other local influences in the area, With the combination of existing models, we can create a novel technology based on ML algorithms that will provide the expected results while maintaining the highest level of accuracy.

They will be helping the industry to take the decisions for the future as well as the present operations and projects going on. From the management perspective, it can give the company a broad perspective on the profits and the other financial aids required. Using AI we can enhance the sight and the growth to the upper levels and can have a longer prediction. Using AI and data science the desired results can be achieved.

7.5 FINANCIAL MODELING WITH THE CURRENT DATA FOR BETTER PERFORMANCE

Financial models are the advanced software developed for the calculations and the commercial workflow and the company. Financial models are the upgraded software for advanced and predictive calculations based on the data available. Data science and data analytics are the sub-applicable technologies in the novel development of the models. The basic financial models were the Excels in which the statistical calculations with the mathematical ones were done and a model is prepared in which entering the few data the entire calculations can be performed shortly. In general, financial modeling involves imitating the characteristics, behavior, and financial implications of a single or multiple financial assets or portfolio of assets, ranging from the loan and the lease and other related agreements to various financial institutions and securities and derivatives. Despite recent research on the academic phase, there is an abundance of material of practical tutorials providing detailed guidelines. The requirement to gather them and make them useful for practical purposes with the advanced technologies (Lukić, 2017).

Also, its integration with AI and ML can be really helpful in the creation of smart models. This model can be customized as per the user company and the work in the petroleum industry. Like the E&P firms, it can manage the billings and the finance as per the work and take accountability for the resources, employees, and other assets of the firm. The digital and the smart system can create the portfolios as per the need. Like if a firm wants the profits and the costs of datasheets and the graphical 3D results, it can generate in a few clicks (Oil and Gas, n.d.). If the firm wants the financial analysis of the human resources, it can also do it on its own and give the results in the format we need. If the financial calculations are needed to be done for the leverages and the loans it can calculate as per the business running and also the predictive timelines up to which it can be solved. Using AI and machines in modelling can result in more efficient calculations and the presentation of data in visual formats that are easier to analyze. Various statistical calculations and processes are involved in the modeling, which needs to be incorporated. As well as the technical aspects are needed to be considered which are also playing a vital role in finance. If the heavier crude is obtained from more viscous and dense locations, both the process conditions and the energy requirements change.

So more economic aspects take place in the larger volumes, so the integrated smart financial program should be developed incorporating the single parameters which are causing the cost to the firm. Thus, it is much more important for the company or firm to keep their accounts and have the data for the same. Jordan and Mitchell (2015) There are programmes available with the firms on which they work, but as technology changes and data security concerns grow, it is necessary to upgrade them to smart technologies such as ML, AI, and the Internet of Things (IoT)-based application for them so that the accuracy and computational results are obtained very correctly, and the data is secured using this system (Jordan & Mitchell, 2015).

Although they are based on the data there are the security measures which are integrated with it giving the companys' data the optimum security. The benefit of using the ML tools is that they can convert and utilize the conventional datasets to the desired results we need. Hoang and Wiegratz (2021) in their study of the ML technologies in financial economics identified the ways to solve problems with ML and the traditional linear aggregation with OLS. Then they focused on the structure of the data available and the construction of the superior and novel measures, reduction in the prediction errors in the economic prediction problems, and the extension of the existing econometric toolsets. Then they focused on the typical price prediction of the real estate. They found that by using ML tools the three aims were achieved. So, in our industry, we can use a similar analogy and algorithms to tailor the gaps in the models and data sets and create novel models in a secure environment with smart technologies to improve the industry and its growth.

7.6 DIGITALIZING THE DISTRIBUTION CHANNEL AND THE END-CUSTOMER THINGS

The end products mainly of the industry are gasoline, diesel, kerosene, and jet aviation fuel. The firms in the downstream sectors are generally involved in the sales directly to the customer. The firms which are also directly involved in the distribution network are an integral part of the system. Converting the established network to the smart digital networks can help in synthesizing the data and the data analytics from the consumer side. Digitalizing the networks helps in maintaining the records and the logistics requirements. The firms in the refinery industries and the commercial distribution of the fuels to the local customers of the common platforms which are fuel stations can be directly connected with the system and with the customer account the data can be automatically retrofitted. So, we can have the individual data for each customer and the volume capacity for the pump for a fixed period, i.e., a month. From that, the distribution server can have the information for the station and can plan the supply management for that automatically. Digital distribution can be regulated with advanced smart technologies where AI and ML can play a vital role. Firms can remind customers to fill up on gasoline or diesel on a regular basis, and the firm can also have a real-time estimate of sales from each fuel station (Hamedi et al., 2009). The revenue created can also be recorded and the prediction can be established for a certain period. In the future with the same system, we can retrofit into different fuels like electric charging stations, biofuels, compressed natural gas, and many others in the research phases with the tailor-made changes required. The stations and the main distribution channel need to be connected with the cloud platform so that real-time information can be communicated. And using AI, the distribution can be planned smoothly. With the power of AI and computational coding, the algorithms can be set up and the secured system can be developed for better performance and observations. If we look up the huge customer base like jet aviation fuel, the same system can be implicated but here the customer base can be smaller than the commercial market. Here the quality and the quantity have to be taken care of and the proper records can help in maintaining the business relations. Other fuels such as kerosene, propylene, and other petrochemicals are also sold to other industries by the refineries for making other products where this petrochemical may be the raw material for them. Digital distribution with ML tools can create more value for the system as ML works majorly on the huge data and the complex algorithms and here the systems and the networks follow the same fashion. Greater opportunities for technology suppliers and developers for the firm, as well as researchers and engineers, to collaborate and develop standardized and cost-effective systems for the industry.

The same can be seen in the natural gas distribution network (compressed natural gas) as well as the piped natural gas used in cooking and commercial applications. The digital distribution network can help pressurize and the metering of the gas (Lancashire, 2014). Similarly, we can connect it to the IoT-based platform and cloud computing so that the server can be connected and have the calculations. Also, it can calibrate the cost to the end user as per the usage in the gas business. The other petrochemical for the refineries can also be monitored and distributed digitally to the industry. Thus using AI- and ML-based digital distribution network and the customer database systems, we can create the novel and advanced model and the system for better operations and performance (Panayiotou, n.d.). The importance of incorporating aspects of the industrial revolution, as well as AI and ML, into each aspect is that next-generation technology will be based on it, and every industry is attempting to adapt it to fit the decorum of the global image.

7.7   SMART SUPPLY CHAIN MANAGEMENT

The intensive and extensive exploitation of the oil and gas industries is causing environmental concerns, and thus sustainability concerns are frequently raised. Sustainability involves the integrated approaches in the economic, social, and environmental units and dimensions of the industry and the business. Any success of the industry depends on supply chain management (Ceptureanu et al. 2018; Chen et al. 2004). The oil and gas industry is such an industry where millions of operations generate a huge quantum of data, which is very hard to scrutinize and analyze on the normal platforms and the trends across that. The oil and gas industry's supply chain is complex and critical in terms of operations. The several key decisions nodes like the purchase of the crude, pricing of the purchases, transportation, logistics, gantry, and the retail sales of the end product. As the final stage arrives for the crude, the complexities in the decision-making increase and the analysis becomes trickier. Areas, where AI and ML can make smooth operations in the supply chains, are the prediction of the market price of crude and the end products. It aids in the decision-making process for price trends and variations. Optimization of crude storage support as well as other logistics support with transportation.

The supply chain involves all the operations we discussed in the upstream, midstream, and downstream sections of the oil and gas industry. Connecting all the domains the centralized system should be integrated and developed so that the domains connected with the IoT can be interpreted and also designed for the separate firms as per the usage, need and application. Chatbots are redefining sources of customer support. The customer engagements can be handled by the bots in the global sight of the oil and gas industry. Using the advanced algorithms the logistics and the mapping facilities are improved further optimization can help it out for better enhancements. Increasing supply chain management efficiency ordering, planning, and control, as well as financial flows in the accounts, payments, and billing sections, are all impacted by SCM and upliftment. With the incorporation of smart technologies, the global oil and gas business can be connected, and business effectiveness and impact can be increased.

The disruption of the chain can have a variety of adverse impacts on the organization including losses in income, market shares, and reputations and even keeping the workers at risk. Cost to respond to supply chain interruptions and can continue operation can incur significant extra expenses over standard operating expenses (Minzner & Johnson 2016). Blockchain technology in the operations can provide a platform to make the energy products digital assets It can reduce the inherent complexity of trading processes by facilitating the process – faster market analysis, lower brokerages, higher time value of information, lower fraud and transactional errors, and many other benefits.

Few topics of the existing studies of the supply chain performance improvements through innovative strategies and the open innovation platforms are like the development of suppliers through management systems able to identify and manage the environmental and social risks involved in their operations. Developing the green supply chain management frameworks for the evaluation of the environmental and social risks involved in the operations. Financial planning, assessment, and payment become easier, as does transition, as a result of developing a smart supply chain and supply chain management for all commodities (Lakhal et al. 2007). The target is to convert the operations and the management of the commodities in an organized and uninterrupted way. The crux should be well defined and clear before developing the system, model, or software. Smart supply chain management will create the communications with the clients; it can also be integrated with chatbots and other AI and ML tools. The requirements and the needs can be achieved and fulfilled by smart supply chain management.

7.8   DEMAND MANAGEMENT WITH COMPUTATIONAL POWER

Demand and supply are the two wheels of the chariot of successful business chains. The demand for the fuel which are the end products for the oil and gas energies are the critical products to maintain the demands as the gasoline and the diesel are the commercial fuels and the demand is variable as per the many factors. Commercial demand management and fuel requirements for public transportation, as well as sir-vehicle jet fuels, must be met. Because the IC engines power the majority of the vehicles, the requirements must be met. Because it is used for the majority of transportation. Managing the demand chain with the advanced technology can give the promising and better results. As per mentioned in the above subsection of the chapter the supply chain management the digital collection of the data from the fuel stations and retrofitting them to the developed model so that using AI we can digitally analyze the trends of the demand and apply the computational powers in the form of ML algorithms for making the process efficient and better result-oriented.

Demands of the fuel requirements for the countries using gasoline and diesel as the fuel can be analyzed and digitally managed as per the usage and the locations of the fuel stations. Since the new technology of the electric vehicle is on its way, some vehicle segments will need to be shifted as a result. The same technology can be applied to the charging station. And then the design and the production of the energy for the charging stations are to be developed. The need and the demand for petrol and other commercial fuels will be the same, products must be changed or the forms may be changed but the need for the management is to be there. The smart and advanced technologies can increase the concept and the design of the system to the optimized level. IoT-based technology and cloud computing can play a major role in the supply chain management and also the demand management of the fuel.

7.9   MANAGING THE VIRTUAL AGENTS AND SUPPLIER SELECTION WITH ML AND DATA SCIENCE

The management of clients and other requirements, as well as the selection of requirements, becomes easier with the help of a large amount of data and the ease of analysis. The data science-integrated approach will help for better data analysis, and AI and ML will carry on and figure out the best possible selection based on the requirements. As the systems are smart and connected on the cloud platform the real-time position and the demand requirements can be noted virtually and the smart supply chain management can track and manage the requirements optimistically. The trends and the growth can be mapped and we can predict the futuristic demand and therefore manage the productions. Data science tools will be playing the greatest role in sorting the data and meeting the selection criteria for the purchase of the product. It can be the process technology, process equipment, maintenance project, and many other intensified technologies. The commissioned technology may not sustain with the advanced systems so there is a need for the replacements and the enhancing efficiency of the process. In the commercial sector, many contractors and E&P companies are working. Managing their technical as well as the economic aspects becomes a somehow drastic job. Applying ML and data science models, as well as sorting and managing data, can greatly assist authorities in making important decisions and tracking falls if necessary to obtain in-depth tasks and system knowledge. On the technical side, we can see the manufacturer and supplier of the equipment or the entire technology, people involved in the commissioning, the head who oversaw the commissioning, and the engineers who operated. All the in-depth data can be compiled under one tree. Maintenance time period zone and the maintenance agencies associated with it can be attached so that the new human resource operation can optimize the scenario and smooth operations can take without any dependency. The fundamentals of AI and ML are that dependency is reduced. While it is impossible to create a system that is completely independent, interdependency can be reduced and the smooth flow of work outputs can occur without any barriers.

The smart system based on AI, ML, and cloud computing technologies will give the acceleration in the management of the commercial and the technical role. Not only in the decision-making for the owners, but also for the operations and the processing, it will play a greater role. Technologies evolve with the decades and are only applied and come to the industry if they are efficient, effective in the economic and performance aspects. The aim is to deliver the economical and best-performance services to the industry; so, only the global acceptance and the technology can achieve the words to the ears. There are many failures and drawbacks in the developing, commissioning, operating, and maintaining. The deal is to balance all of this and provide the best results possible. There are lots of opportunities for the researcher and the R&D section to make the applied research viable in the expected parameters. As lots of money and utilities are employed to make the process effective and achieve maximum efficiency. So, using AI and ML, as well as simulations, we can actually perform the process and see how it performs, and we can also use AI and ML to judge the system's economic conditions.

REFERENCES

Al-Fattah, S.M. (2019). Artificial intelligence approach for modeling and forecasting oil-price volatility. *SPE Reservoir Evaluation & Engineering 22*, 817–826. https://doi.org/10.2118/195584-PA

Boz MF, Arslan M (2017). Analysis of the factors affecting the capital structure of oil exploration and production companies: Comparative analysis of TP and the five major oil exploration and production companies in the world. *Journal of Business Research* – Turk, 9(2),12–231. https://doi.org/10.20491/isarder.2017.269

Ceptureanu, S.I., Ceptureanu, E.G., Olaru, M., Popescu, D.I. (2018) An exploratory study on knowledge management process barriers in the oil industry. *Energies*, 11(8), 1977. https://doi.org/10.3390/en11081977

Chen, I.J., Paulraj, A. (2004). Towards a theory of supply chain management: the constructs and measurements. Journal of Operations Management, 22(2), 119–150.

Ding, Y. (2018). A novel decompose-ensemble methodology with AIC-ANN approach forcrude oil forecasting. *Energy*, *154*, 328–336.

Hamedi, M., Zanjirani, R., & Moattar, M. (2009). A distribution planning model for natural gas supply chain: A case study. *37*, 799–812. https://doi.org/10.1016/j.enpol.2008.10.030

Jordan, M. I., & Mitchell, T. M. (2015). Machine learning: Trends, perspectives, and prospects. *349*(6245), 255–260. DOI: 10.1126/science.aaa8415

Kumar, R. (2017). Value drivers in oil companies: An application of variance based structure equation model. *13*(1), 31–52. https://doi.org/10.7903/cmr.16165

Lakhal, S.Y., H'Mida, S., Islam, M. R. (2007). Green supply chain parameters for a Canadian petroleum refinery company. *International Journal of Environmental Technology and Management*, 7, 56–67.

Lukić, Z. (2017). The art of company financial modelling. *Croatian Operational Research Review*, 8, 409–427. https://doi.org/10.17535/crorr.2017.0026

Minzner Conley, A., & Johnson, L. (2016, June). No Weak Links in Your Supply Chain. In ASSE Professional Development Conference and Exposition. OnePetro.

Panas, E., Vassilia N. (2000) Are Oil Markets Chaotic? A Non-linear Dynamic Analysis. *Energy Economics 22*(5), 549–568.

Panayiotou, N. A. (n.d.). Applying the industry 4.0 in a smart gas grid: The Greek gas distribution network case. 180–184.

Yu, L., Wang, S., Lai, K. K., 2008. Forecasting crude oil price with an EMD-based network ensemble learning paradigm. *Energy Economics*, *30*(5), 2623–2635.

Yu, L., Yang, Z., Ling, T. (2017) Ensemble forecasting for complex time series using sparse representation and neural networks. *Journal of Forecasting, 36*(2), 122–138.

Chapter 8

Market and trading in oil and gas (petroleum) industry

8.1 INTRODUCTION

Oil and gas (petroleum) have remained one of the essential sources of energy for many years. Oil and gas, as well as their products, are consumed by all countries. Both producers and consumers are interested in oil and gas pricing and derivatives (Amberg & Fogarassy, 2019). The price of oil and gas influences the level of costs in all sectors of production. Because many countries' economies are built on oil and gas production and trading in oil and gas products, predicting oil and gas prices is essential. It is important to note that oil and gas prices directly impact some sectors of the economy (Dash & Maitra, 2021). Oil and gas prices affect political and economic processes that influence the stock value of oil and gas businesses, the inflation rate in oil- and gas-importing countries, and economic growth (Batabyal & Killins, 2021). In 2014–2019, the average annual volume of oil consumption was over 4.2 billion tons, which increased 54% from 1974 to 1979. As a result, the average annual rise in oil and gas consumption since the oil shock was 1%. At the same time, following the 1973–1983 economic crisis, oil consumption increased consistently until the 2008 crisis. Significant and unexpected variations in oil and gas prices, on the other extreme, are widely believed to have a negative influence on the well-being of both oil- and gas-importing countries and oil and gas producing countries. In 2018, shale gas accounted for over half of all natural gas manufacturing in the United States (Solarin et al., 2020). Total primary energy supply analysts predict oil at 32%, coal at 29%, natural gas at 23%, biofuel and waste at 10%, nuclear at 5%, hydro at 2%, and others at 0.15%. The third-largest energy source in the world is natural gas. As a result, more wells are being drilled, and new natural gas reservoirs are being searched to supply this need.

8.2 OIL AND GAS (PETROLEUM) INDUSTRY MARKET DYNAMICS

Knowledge is a collection of information, data, experience, and expert opinion that helps assess and integrate new information and experience. Sharing the information is crucial to a company's profit, which leads to implementing the information faster in every sector of the company, leading to a profit and improving the industry's performance and competitiveness. The petroleum industry is one of the major and the primary force behind several others, including transportation. Global industries are exposed to working with people from many backgrounds, cultures, and settings due to

DOI: 10.1201/9781003279532-8

undertaking such enormous projects in different world regions. One of the really difficult responsibilities in the oil and gas companies is data and information management
(Pollo et al., 2018). Engineers are frequently distracted by enormous amounts of data,
causing them to ignore crucial information that might otherwise help them better understand the reservoir. As a result, petroleum data analysis has become increasingly
popular.

Forecasting petroleum production time series is a critical activity that supports oil
refinery output scheduling (K I & O H, 2014). Forecasting, on the other hand, necessitates determining whether a time series is chaotic. In this study, K I and O H (2014)
looked at chaos analysis-based forecasting of diesel and petrol production time series. Based on Lyapunov exponents, the differential entropy approach determines the
ideal embedding dimension values and time lag in chaos analysis. Fuzzy "IF-THEN"
rules built based on time-series analysis using fuzzy clustering are used to forecast the
amount of future petroleum production (Vilela et al., 2018). The adequacy of the constructed nonlinear forecasting model is illustrated using the prediction results.

This article uses an adaptive network-based fuzzy inference system (ANFIS)
(Karaboga & Kaya, 2018; Qanbari et al., 2013) to predict natural gas consumption.
Forecasting future natural consumption of gas can assist Statesmen in making more
informed decisions. Numerous variables influence gas consumption, but gross domestic product (GDP) and population, two input variables have been chosen. From 1993
through 2012, data on input variables and output variables (gas consumption) were
collected. Fuzzy models, ANFIS, are used in this research, and the outcomes and
errors of each model are explored. The mean absolute percentage error (MAPE) com-
pares all ANFIS outputs to the real output. For forecasting gas consumption, the best
model with the lowest MAPE is chosen. The testing for many trained fuzzy models
and the results and errors of every model is made in this (Qanbari et al., 2013) research.
In chaos analytics, the ideal values of embed dimension and lag time are determined.
Because it has the lowest MAPE, the Gaussian curve built-in membership function
(Wang et al., 2020) and three different variables, the model is the best among the other
autoregressive model is used to forecast population and GDP from 2013 to 2020. They
are supplied as inputs into the best ANFIS model, and as outputs from 2013 to 2020,
the gas consumption values are calculated. It is undeniable that natural gas consumption is rapidly rising. As a result, the authorities should take it seriously.

Chithra Chakra et al. (2013) propose a unique neural technique for estimating oil
output utilizing a higher-order neural network (HONN). To demonstrate the forecasting capacity of HONN models, two case studies were conducted using the Cambay
basin, Gujarat, India oil field's data. According to the simulation results, HONN has
strong potential with few petroleum reservoirs for oil production projection input parameters (Pattanayak et al., 2020). In the two scenarios studied, the HONN approach
used to forecast oil output generated MAPEs of 13.86 and 15.13. Selection of optimal
input variables and noise reduction are the two preprocessing steps to help achieve the
MAPE. Estimating using the few input parameters provided indicates that HONN
has a high probability of overcoming this constraint. For complex input patterns, the
computational cost is reduced using the high combo of the input items, which gives a
rapid output.

To forecast oil prices machine learning (ML) algorithms based on the standard, regression analysis, and enhanced linear regression are applied, and the factors

affecting the oil price are proposed in this excerpt (An et al., 2019b). Price forecasting can be done using factors such as the U.S. key rate, the U.S. dollar index, the S and P 500 index, the volatility index, and the U.S. consumer price index. We may conclude that oil prices in 2019–2022 will have a modestly increasing trend and will be generally stable after examining the findings and comparing the model's accuracy first. Brent's price had dropped to a 17-month low at the time of the drop in June 2012. The reason for this was low demand for oil futures, which was prompted by skewed statistics on the labor market status in the United States. As an alternate source of investment (An et al., 2019a) expected that gold would fall on oil at the beginning of our research. However, it was not confirmed. The popularity of precious metals investing has no impact on the company's investments in oil stocks. It concluded that the oil prices will increase slightly and will be relatively constant from 2019 to 2022.

The electric company can use precise demand forecasts to make unit commitment choices, reduce rotating reserve capacity, organize device maintenance programs more effectively, lower generation costs, and increase power system reliability (Parth Manoj & Pravinchandra Shah, 2014). Short-term evaluation measures are attempted using a fuzzy logic technique in this article (Parth Manoj & Pravinchandra Shah, 2014). For short-term load forecasting, temperature, time, and a same past day load are employed as not dependent variables. Using Mamdani's implication, a fuzzy rule base is created based on the time, temperature, and similar previous day load, then used for load forecasting in the short term. Load forecasting is done with a margin of error of +2.695659% and −1.884780% using temperature, time, and similar prior day's load as inputs and accessible data to create a fuzzy logic rule basis. Furthermore, because it is based on basic "IF-THEN" statements, the fuzzy logic technique is straightforward for understanding the forecast. It also helps in spinning reserve capacity reduction, unit commitment decisions, and device maintenance schedules.

Considering conventional methods, the production of waterflooding reservoirs is difficult to estimate using a production forecasting model based on artificial neural networks (Negash & Yaw, 2020). Some examples were shown by Negash and Yaw (2020). To increase the prediction effect, a methodology involving physics-based feature extraction was presented for fluid production forecasting. The Bayesian regularization model was used as the model's algorithm for training. Even though it takes longer, this method can more accurately generalize data sets on oil, gas, and water production. The model was evaluated by calculating the determination coefficient and mean square error, as well as creating an error distribution histogram and a cross-plot between simulation and verification data. During the validation data of the chosen neural network design, it was revealed that one trained with Bayesian normalization had the lowest mean square error as well as the highest regression coefficients. This technique takes longer, but it can generalize noisy datasets like gas, oil, and water production rates. The results reveal that the model's flaws are most likely attributable to measurement inaccuracies.

Forecasting approaches fuzzy rule-based time series have gained prominence in recent years (Wulandari et al., 2020). In this study, Wulandari et al. (2020) present a new method for defining the universe of converse, historical data variation, and the partitioning state. The discourse universe was discussed early on, then calculated the primary value to determine how many intervals should be employed with changing historical data. Second, breaking down the primary intervals into multiple sub-intervals.

According to the empirical investigation, the fuzzy number became closer to the crisp value as a result of the sub-interval. It results in a higher forecasting value. For simulation, data from Indonesia's annual petroleum output were used. The forecasting outcomes and error value methods are compared to those of earlier approaches. The improvements produce better forecasting outcomes than previous techniques with reduced average forecasting error and mean squared error (Naderi et al., 2019). From 1996 to 2017, a unique algorithm method was used to forecast petroleum data gathered from *bps.co.id*. The first-order fuzzy time series are compared with an inaccuracy of 0.033% using frequency density partitioning. The first-order error varies from 2% to 3% when using first-order fuzzy time series based on the frequency density partitioning technique. Computers (and robots) now have the capacity to have intuition thanks to recent advancements in neural networks.

This research presents a new flower pollination strategy for NN training to develop a model for the Organization of Petroleum Exporting Countries (OPEC) (Chiroma et al., 2016; Ji et al., 2019) to forecast petroleum consumption with a surprising balance between consistency and exploration. The stated method is contrasted with well-known metaheuristic methods. The outcomes reveal that the newly stated method surpasses algorithms' accuracy and convergence speed existing for OPEC petroleum consumption forecasts. Our proposed method can be a useful tool for OPEC lead persons and other worldwide oil-related organizations in projecting OPEC petroleum consumption. Compared to the methods previously addressed in the literature, this work presented an alternative way of neural network training with increased convergence speed, accuracy, and resilience. The approach provided in our research for estimating the consumption of OPEC oil can assist the OPEC administration in monitoring, controlling, designing, modifying, and implementing OPEC petroleum consumption regulations more efficiently. The proposed algorithm's performance should be tweaked to include the capacity to explore oceanographic and meteorological, and big data datasets of oil and gas.

8.3   DATA ANALYSIS AND MARKET FORECASTING OF PRICES FROM THE RAW MATERIAL PRODUCED IN THE OIL AND GAS (PETROLEUM) INDUSTRY

The approach of constructing a consensus forecast is now used to determine the forecast price for oil. This strategy is based on oil market forecasts from the most significant players, investment banks, and international, economic, and financial organizations, including OPEC, International Energy Agency, IHS Global Insight, International Monetary Fund, World Bank, and Raiffeisen Bank (Basher et al., 2018). The oil price is one of the most important elements defining the Russian budget in terms of revenues. Each prediction approach has its own set of disadvantages; due to the closed nature of the methods utilized, it is impossible to assess the degree of the prediction inaccuracy. Using consensus and results from several sources can lead to "inheriting" the shortcomings of the prognostic source. When initial estimates are based on certain assumptions and hypotheses, consensus prediction essentially eliminates the outcome, altering the original value of predictions acquired from other sources. According to an analysis of the practice of developing predictive estimations and forecasting methods employed by many research institutions, governments, corporations, and pits, the most popular machine learning (ML) methodologies are based on econometric forecasting methods. In this approach, it is proposed to apply an ML-based prediction method as an alternative to consensus forecasting.

Aims to estimate Nigeria's petroleum (thousands of barrels per year) consumption (Folorunso & Vincent, 2019). The consumption of petroleum was predicted using random forest regression (RFR), linear regression, and autoregressive integrated moving average (ARIMA) models. Coefficient of determination, MAPE, root mean square error (RMSE), and mean absolute error (MAE) metrics are the models used to evaluate the prediction accuracy. Folorunso & Vincent (2019) propose linear regression (LR) and RFR for estimating consuming petroleum in Nigeria, and the results were compared to the ARIMA in this research (Okey Onoh & Peter Eze, 2019). The four processes that the ML models for time-series forecasting went through were collecting the data, preprocessing the dataset (Differencing), training and learning, and forecasting the test. For evaluating the effectiveness of the ML models based on the coefficient of determination, RMSE, MAPE, and MAE, an empirical investigation comparing the performance of LR and RFR with ARIMA (1,1,1) is being conducted. In the yearly prediction of Nigerian petroleum consumption, the results reveal that LR excels in RFR and ARIMA forecasting methodologies. The accuracy of the forecast can be increased by fine-tuning the parameters of the ML models.

One of the most important ideas in finance is stock price prediction (Ghanbari & Jamshidi, 2019). Traders and investors may be able to predict stock values owing to ML more correctly. The dependent variable in Ghanbari and Jamshidi (2019) paper is the closing price. In contrast, the independent variables are the total index, volume, last price, first price, opening price, today's low, today's high, exchange rate, Brent index, Tehran Stock Exchange, and WTI index. Seven different ML techniques are used to estimate stock prices, including boosted tree, decision forest, Bayesian linear, ensemble regression, support vector, and neural networks. Each algorithm's performance was measured using two metrics: mean square error and mean absolute error. The Bayesian linear regression showed a good performance in predicting the stock price in the oil and gas industry on the Tehran Stock Exchange (Shirvani & Volchenkov, 2019). Machine learning algorithms are implemented in the counter (OTC) market to study the oil and gas business's stock behavior. It would be advantageous for the Tehran Stock Exchange to use a segmentation category of machine learning techniques to classify equities in the oil and gas business with varying activity and riskiness.

This paper describes the current advances achieved by researchers, particularly in the sector of estimating shale gas production performance using ML-based models, such as support vector machine (SVM), decision tree model (Syed et al., 2021), random forest regression, gradient boosting for regression tree (Al-Qutami et al., 2017) and Gradient boosted machine (GBM) (Song & Zhou, 2019). In addition, this (Syed et al., 2021) explains equations, input parameters, and formations regarded as significant parameters in the construction of smart shale gas models. Eagle Ford Shale, Marcellus Shale, and Bakken Shale are North America's largest shale reservoirs discussed in the paper. The research on using artificial intelligence (AI) and ML-based algorithms producing characteristics and parameterization of oil and gas wells in three major U.S. coal seams were comprehensively examined in this work. The application of ML and AI shale gas reservoirs was conducted in intensive research in 2018. Since 2018, extensive research has been undertaken on ML and AI (Kshirsagar & Shah, 2021) in shale gas reservoirs. It is concluded that, with input from significant sample data drilled in individual formations, almost all established models can forecast the production behavior of oil and gas wells. With improved precision, accuracy, and efficiency, the dependable model can monitor the outcome of shale gas deposits in terms of production.

Random forest, lightgun, XGBoost, stochastic real-valued, multilayer perceptron neural network, and Super Learner are the six ML models implemented in this paper to predict dead oil viscosity (Hadavimoghaddam et al., 2021). These models were developed and tested using over 2000 pressure–volume–temperature data (Sinha et al., 2020). Hadavimoghaddam et al. (2021) examine the effectiveness of various functional forms that have been employed to model dead oil viscosity in the literature. The results demonstrate that the functional form (gAPI, T) performs well and that some correlating factors may not be required. Additionally, Super Learner excelled in other ML methods and common measure analysis correlations. It has the potential to increase the precision of crude oil viscosity modeling where some data is lacking. The results support the suggested Super Learner model's performance in estimating dead oil viscosity. The results indicate that the most basic functional form for calculating viscosity is sufficient to produce statistically valid and acceptable results. It means that extra correlation factors may not be required to increase a model's performance.

Present a deep learning methodology eligible for overcoming the drawbacks of traditional predicting methods while also delivering accurate forecasts (Sagheer & Kotb, 2019). A deep long-term memory structure is an RNN-based expansion. Two petroleum industry case studies are reviewed, with productivity data from two genuine oilfields used. The empirical results indicate that other common approaches are outperformed by the suggested deep long-term memory model using various measurement criteria. The presented method is a deep long-term memory, or deep recurrent network long short-term memory architecture (LSTM). The study demonstrates that layering additional LSTM layers ensures that the limitations of shallow neural network designs are restored, especially when utilizing time-series datasets with long intervals. Moreover, the suggested deep model could represent the nonlinear connection between systems outputs and inputs, which is particularly useful when the petroleum time-series data is diverse, complicated, and missing portions. Notably, in the case studies reported in this work, the suggested model outperforms its deep RNN and deep gated recurrent unit (DGRU) equivalents.

The oil production is initially divided into training and test sets (Liu et al., 2020). EEMD decomposes the test set data and adds it to the training set, yielding numerous intrinsic mode functions. The curve and mean similarity computed via dynamic temporal warping is used to assess the stability of intrinsic mode functions. A genetic algorithm is used to identify the best LSTM hyper-parameters. Two genuine oilfields are used to test the proposed model in China for verification and evaluation. The presented method is capable of producing near-perfect production forecasts, according to empirical results. Genetic algorithm is used to find the best LSTM hyper-parameters and architecture. Instead of using the usual "divide and conquer" technique, the EEMD-LSTM can predict oil production movements immediately (Peng et al., 2021). The proposed approach can provide excellent prediction of oil production since the fluctuations of the oil production series and context information are well established. The suggested EEMD-LSTM outperformed the EEMD-SVM and EEMD-ANN according to the empirical results. In both the case study of SL and JD oilfields, the EEMD-LSTM model has the lowest errors and the highest determination coefficient, demonstrating that the suggested model may provide almost flawless oil production forecasting in time series.

The outcomes of predicting oil, water, and gas production using an approach that combines two AI models in a Colombian oil field are presented (Ruiz et al., 2019). A unique data mining technique, comprising a data imputation strategy, is constructed by combining artificial neural networks (ANN) and fuzzy logic. The fuzzy logic tool assesses which variables or characteristics are the most valuable to include in each well's production model. After the data mining process, prediction models such as ANN and fuzzy inference systems are constructed. Fuzzy inference system models can forecast particular actions, while ANN models can forecast average behavior. With only a few iterative steps, the combined use of both tools allows for improved well-behavior prediction until a certain degree of accuracy is obtained. The proposed imputation procedure is critical for correcting faulty items or filling in voids models for a typical oil production field were discovered using performance data. For each well model, a random search is performed to identify the significant variables that will be used as model input (regressor). The random search is performed a finite number of times. Combining each estimated outcome and acquiring the most accurate fluid phase production projection possible above ANN and fuzzy inference systems is significant and effective.

8.4   VALUATION OF DERIVATIVES OR ASSETS OF OIL AND GAS (PETROLEUM) INDUSTRY

The viscosity of crude oil is determined experimentally on subsurface or surface samples at reservoir temperature and pressure (Alhammadi et al., 2017). However, this is generally costly and time-consuming and requires a strong technical specialization. In this regard, over the last few decades, a vast number of empirical and semi-empirical relationships have been created, primarily based on the corresponding equation of state to forecast crude oil viscosity. Most of the correlations mentioned have been created for a specific region, and thus if they are applied to other areas, they will produce incorrect findings. India is an agricultural-based country that is looking to boost its economy by increasing fuel subsidies. As a result, diesel is less expensive than petroleum. It is possible that this is not the case in some other countries' economies. Furthermore, each country's importing patterns will have an impact on petrol prices. A country with a lower additional tax charge is cheaper for both petroleum and diesel. Oil trading firms have changed the additional tax submitted to some state taxes. As a result, oil-based commodities costs have risen in states like Assam, Mumbai, Maharashtra, Karnataka, Gujrat, and West Bengal. The 2002 Irrecoverable Taxes Compensation Scheme should be reviewed to account for the expenses in states where permanent duties have been decamped.

AI techniques are discussed in this study for a better plan, determining rock reservoir parameters, optimizing the drilling and production facilities (Solanki et al., 2021). Reservoir engineers can build a solid plan for developing the reservoir and successfully control hydrocarbon recovery with exact knowledge of permeability and porosity.

The model of AI techniques has been studied to estimate the single-phase fluid permeability by considering the various log data types. The solution to specific challenges that arose during the permeability determination in the real experiment is also mentioned Solanki et al. (2021) concluded that the numerous oil and gas (petroleum) industry segments use AI techniques (Kshirsagar, 2018). Its methodology has a considerable impact on discrete applications, resulting in risk reduction, cost reduction, and the resolution of complex problems. AI algorithms were utilized to forecast relative permeability curves identical to those found in the actual experiment. It results in precise permeability data which is not affected during the tests, and it eliminates the issues associated with the traditional approach to determining permeability data. Using AI approaches, the oil and gas (petroleum) industry has made significant progress in the production, drilling, EOR, and reservoir sectors. AI approaches are reliable, but the major drawback is the cost. The technique of fuzzy logic was utilized to create EOR models based on the available data (Kumar & Barua, 2021). Expert systems like SAVA and NADA can be useful in determining the optimal EOR technique and assessing the recoverable oil amount using EOR technologies.

Aims to determine the essential success elements for petroleum company information centers (Hawash et al., 2020). The relevance of combining these factors with the use of AI and ML technology in oil firms is to obtain better output in responding to terrible risks and surviving. It is also critical to the company's performance and survival and enables information exchange in the face of present and future difficulties. The significance of crucial success elements for information centers in ensuring the company's survival and enhancing its performance can be used to justify their importance, which necessitates experience, context understanding, and familiarity with the implementation team. To meet present and future problems and create a vision for transformation and progress that is logical and consistent, it is vital to determine the critical success criteria, their significance in the company's existence in the field, and how it is matched with the plan of the business. Officials must have a force to control all critical issues to succeed in executing these variables. However, due to high costs, most oil corporations have opposed using AI and ML.

Abolfazli et al. (2014) discuss how time-series and artificial neural networks can improve energy forecasting in the rail transportation sector. The optimal input variables for the integrated ANN model are determined using autocorrelation and partial autocorrelation functions in an integrated ANN model. The suggested ANN uses partial autocorrelation function and autocorrelation function, retrieved from time-series information, to pick applicable inputs for ANN. The two regressive auto models are compared using analysis of variance techniques to evaluate the ANN results (Yildirim et al., 2019). Total weekly consumption of the crude oil in Iran railway transportation is used to build and compare time-series and ANN models from January 2009 to October 2011. It is estimated that ANN produces excellent results and may be further utilized for prediction. Another study shows partial autocorrelation function and autocorrelation function analysis to determine time-series modeling using the ANN inputs. The integrated ANN in this work can deal with data correlation, autocorrelation, complexity, and nonlinearity due to its mechanism.

Forecasting oil production is vital in petroleum engineering because it helps engineers manage petroleum reservoirs (Al-Shabandar et al., 2021). However, with the rise of digital oil big data, accurate production forecasting is becoming increasingly challenging. Despite a large amount of work in the literature on ML in the oil and gas sector, existing prediction approaches can only represent the complex elements of time-series analysis. The model suggested by Al-Shabandar et al. (2021) has a simple design and can track time-series datasets with lengthy intervals. The new technique was compared to other traditional approaches to assess the scalability of our model. The proposed model outperforms existing techniques, according to comprehensive empirical results. A novel deep recurrent neural network was built in this study. Petroleum engineers can use the proposed model to track the characteristics of producing oil wells. The model is based on the gated recurrent unit (GRU) recurrent neural network's deep structure. Because the suggested model outperforms other common models, our DGRU model can be used to analyze the long-term dependencies of a complicated time-series dataset. Deep generative models, which can derive high-level representations from high-dimensional unstructured data, could be implemented (Gurwicz, 2020).

Knowledge-based systems are AI tools that store and apply professional opinions, methods, and rules to provide exact system results (Ghallab et al., 2013). As part of developing an expert system, data from oil wells was analyzed, and memberships of petroleum oil fields were reordered. Fuzzy petroleum prediction has been built as a well-versed system in this research. The critical data was gathered from a variety of sources. Temperature, pressure, crude oil density, gas density, and gravity are the five parameters to predict petroleum prices. The expert system covers 30 wells in the Daqing oil field for petroleum prediction. In petroleum prediction, the fuzzy expert system is utilized to carry out the prediction process and assess the results accurately. The capacity of fuzzy systems to generate uncertain domains, such as the petroleum domain, is one of these complex subjects. The main purpose of the prediction is to encourage engineers to grow the oil and gas industries. Fuzzy systems use petroleum datasets, oilfield classification, petroleum presence prediction, accuracy, and evaluation. The fuzzy petroleum prediction system produces close to exact results for the measured parameters of the wells. In the petroleum industry, a variety of expert systems can be deployed.

Recent advances in neural networks have enabled computers (and machines) to come up with a reasonable solution to a problem that is unreasonably difficult to solve using regular logical methods (Ali, 1994). Even the input data is imprecise and noisy, and neural networks can learn complicated nonlinear relationships. Pattern recognition, noisy data categorization, nonlinear feature identification, market prediction, and process modeling have benefited from neural networks. Because of these characteristics, neural network technology is well suited to solving problems in the petroleum sector. The goal of this study is to review the areas of petroleum technology where artificial neural networks have indeed been effectively applied and also to suggest new applications. Only a few examples include seismic pattern recognition, permeability predictions, recognition of sandstone lithofacies, drill bit diagnostic and analysis, or gas well production enhancement. The use of neural networks technology could assist with well performance analysis, prediction, optimization, integrated reservoir characterization, and portfolio management.

The neural network models are built using a systematic workflow in this study (Al-Bulushi et al., 2012). The method included various design difficulties for generating neural network models, particularly when constructing the model's structure. We evaluate the data's statistics for relevance. The results were compared to those of other regression models using the calculation of the impact of the input variable. Petroleum (hydrocarbons) is a critical source of energy all over the world. One of the aims of the petroleum industry while identifying an oilfield is to obtain an accurate estimation of the volume of the hydrocarbon in situ before any investment is committed to development and manufacturing. Water saturation is one of the most significant criteria to consider while determining reserves. This approach addresses a variety of design difficulties that arise during the construction of ANNs. The workflow created an ANN model to predict water saturation in a petroleum field. Only specific wells have the core data that allows for accurate measurement of water saturation. The ANN was used to solve the complex nonlinear connection between wireline logs and core saturation. In addition, the ANN outperformed traditional statistical tools (such as multiple regressions) (Table 8.1).

Table 8.1 Comparative analysis of the studies

Ref.	Year	Contributions	Results
Folorunso and Vincent (2019)	2019	This research aims to estimate Nigeria's petroleum (thousands of barrels per year) consumption. The consumption of petroleum was predicted using RFR, LR, and ARIMA models. Coefficient of determination, RMSE, MAPE, and MAE metrics are the models used to evaluate the prediction accuracy.	The yearly prediction of Nigerian petroleum consumption results reveals that LR excels in RFR and ARIMA forecasting methodologies. The accuracy of the forecast can be increased by fine-tuning the parameters of the ML models.
Folorunso and Vincent (2019)	2019	In the yearly prediction of Nigerian petroleum consumption, the results reveal that LR excels in RFR and ARIMA forecasting methodologies. The accuracy of the forecast can be increased by fine-tuning the parameters of the ML models.	To meet present and future problems and create a vision for transformation and progress that is logical and consistent, it is vital to determine the critical success criteria, their significance in the company's existence in the field, and how it is matched with the plan of the business.
An et al. (2019b)	2019	An ML algorithm based on the standard linear regression and modified linear regression is used to predict the oil prices, and the factors affecting the oil price are proposed in this paper.	The conclusion is that oil prices in 2019–2022 will have a modestly increasing trend and will be generally stable after examining the findings and comparing the models of ML algorithms' accuracy first. It concluded that the oil prices will increase slightly and will be relatively constant from 2019 to 2022.

(Continued)

Table 8.1 (Continued) Comparative analysis of the studies

Ref.	Year	Contributions	Results
Negash and Yaw (2020)	2020	Considering conventional methods a production forecasting model based on artificial neural networks has problems predicting the production of waterflooding reservoirs, and some examples were shown. To increase the prediction effect, a methodology involving physics-based feature extraction was presented for fluid production forecasting.	It was discovered that a neural network architecture trained using Bayesian regularization yielded the minimum mean square error and the highest coefficient of determination. This technique takes longer, but it can generalize noisy datasets like gas, oil, and water production rates. The results reveal that the model's flaws are most likely attributable to measurement inaccuracies.
K I and O H (2014)	2014	Fuzzy models like ANFIS are used in this research, and the outcomes and errors of each model are explored. MAPE compares all ANFIS outputs to the real output. For forecasting gas consumption, the best model with the lowest MAPE is chosen. The testing for many trained fuzzy models and the results and errors of every model is made.	The autoregressive model is used to forecast population and GDP from 2013 to 2020. They are supplied as inputs into the best ANFIS model, and as outputs from 2013 to 2020, the values of gas consumption are calculated. It is undeniable that natural gas consumption is rapidly rising. As a result, the authorities should take it seriously.
Al-Shabandar et al. (2021)	2021	The model suggested has a simple design and can track time-series datasets with lengthy intervals. The new technique was compared to other traditional approaches to assess the robustness of our model. The model is based on the GRU recurrent neural network's deep structure.	The DGRU model which is proposed in this paper outperforms existing techniques, according to comprehensive empirical results. Our DGRU model can be used to analyze the long-term dependencies of a complicated time-series dataset. Deep generative models, which can derive high-level representations from high-dimensional unstructured data, could be implemented.

REFERENCES

Abolfazli, H., Asadzadeh, S. M., Nazari-Shirkouhi, S., Asadzadeh, S. M., & Rezaie, K. (2014). Forecasting rail transport petroleum consumption using an integrated model of autocorrelation functions-artificial neural network. *Acta Polytechnica Hungarica*, *11*(2), 203–214. https://doi.org/10.12700/aph.11.02.2014.02.12

Al-Bulushi, N. I., King, P. R., Blunt, M. J., & Kraaijveld, M. (2012). Artificial neural networks workflow and its application in the petroleum industry. *Neural Computing and Applications*, 21(3), 409–421. https://doi.org/10.1007/s00521-010-0501-6

Al-Shabandar, R., Jaddoa, A., Liatsis, P., & Hussain, A. J. (2021). A deep gated recurrent neural network for petroleum production forecasting. *Machine Learning with Applications*, 3, 100013. https://doi.org/10.1016/j.mlwa.2020.100013

Al-Qutami, T. A., Ibrahim, R., & Ismail, I. (2017). Hybrid neural network and regression tree ensemble pruned by simulated annealing for virtual flow metering application. *Proceedings of the 2017 IEEE International Conference on Signal and Image Processing Applications, ICSIPA 2017*, 304–309. https://doi.org/10.1109/ICSIPA.2017.8120626

Alhammadi, A. M., AlRatrout, A., Singh, K., Bijeljic, B., & Blunt, M. J. (2017). In situ characterization of mixed-wettability in a reservoir rock at subsurface conditions. *Scientific Reports*, 7(1), 1–9. https://doi.org/10.1038/s41598-017-10992-w

Ali, J. K. (1994, March 15). Neural networks: A new tool for the petroleum industry? *All Days*. https://doi.org/10.2118/27561-MS

Amberg, N., & Fogarassy, C. (2019). Green consumer behavior in the cosmetics market. resources, 8(3), 137. https://doi.org/10.3390/RESOURCES8030137

An, J., Mikhaylov, A., & Moiseev, N. (2019a). International journal of energy economics and policy oil price predictors: machine learning approach. *International Journal of Energy Economics and Policy*, 9. https://doi.org/10.32479/ijeep.7597

An, J., Mikhaylov, A., & Moiseev, N. (2019b). Oil price predictors: Machine learning approach. *International Journal of Energy Economics and Policy*, 9(5), 1–6. https://doi.org/10.32479/ijeep.7597

Basher, S. A., Haug, A. A., & Sadorsky, P. (2018). The impact of oil-market shocks on stock returns in major oil-exporting countries. *Journal of International Money and Finance*, 86, 264–280. https://doi.org/10.1016/J.JIMONFIN.2018.05.003

Batabyal, S., & Killins, R. (2021). The influence of oil prices on equity returns of Canadian energy firms. *Journal of Risk and Financial Management*, 14(5), 226. https://doi.org/10.3390/JRFM14050226

Chiroma, H., Khan, A., Abubakar, A. I., Saadi, Y., Hamza, M. F., Shuib, L., Gital, A. Y., & Herawan, T. (2016). A new approach for forecasting OPEC petroleum consumption based on neural network train by using flower pollination algorithm. *Applied Soft Computing*, 48, 50–58. https://doi.org/10.1016/J.ASOC.2016.06.038

Chithra Chakra, N., Song, K. Y., Gupta, M. M., & Saraf, D. N. (2013). An innovative neural forecast of cumulative oil production from a petroleum reservoir employing higher-order neural networks (HONNs). *Journal of Petroleum Science and Engineering*, 106, 18–33. https://doi.org/10.1016/j.petrol.2013.03.004

Dash, S. R., & Maitra, D. (2021). Do oil and gas prices influence economic policy uncertainty differently: Multi-country evidence using time-frequency approach. *The Quarterly Review of Economics and Finance*, 81, 397–420. https://doi.org/10.1016/J.QREF.2021.06.012

Folorunso, O., & Vincent, O. R. (2019). Stock price trend prediction using support vector machine and coral reef, 68–87.

Ghallab, S. A., Badr, N., Hashem, M., Salem, A. B., & Tolba, M. F. (2013). A fuzzy expert system for petroleum prediction. *Recent Advances in Computer Science and Applications*, 70–82.

Ghanbari, A. M., & Jamshidi, H. (2019). Machine learning application in stock price prediction: applied to the active firms in oil and gas industry in Tehran stock exchange. *Petroleum Business Review*, 3(2), 29–41. https://doi.org/10.22050/PBR.2019.112803

Gurwicz, A. (2020). *Deep Generative Models for Reservoir Data: An Application in Smart Wells*, PhD Thesis. Rio de Janeiro: Pontifical Catholic University of Rio de Janeiro (PUC-Rio).

Hadavimoghaddam, F., Ostadhassan, M., Heidaryan, E., Sadri, M. A., Chapanova, I., Popov, E., Cheremisin, A., & Rafieepour, S. (2021). Prediction of dead oil viscosity: machine learning vs. classical correlations. *Energies*, 14(4), 930. https://doi.org/10.3390/en14040930

Hawash, B., Mokhtar, U. A., Yusof, Z. M., & Mukred, M. (2020). The adoption of electronic records management system (ERMS) in the Yemeni oil and gas sector: Influencing factors. *Records Management Journal*, *30*(1), 1–22. https://doi.org/10.1108/RMJ-03-2019-0010/FULL/HTML

Jabbarova, K. I., & Huseynov, O. H. (2014). Forecasting petroleum production using chaos time series analysis and fuzzy clustering. *ICTACT Journal on Soft Computing*, *4*(4), 791–795. https://doi.org/10.21917/ijsc.2014.0113

Ji, Q., Zhang, H. Y., & Zhang, D. (2019). The impact of OPEC on East Asian oil import security: A multidimensional analysis. *Energy Policy*, *126*, 99–107. https://doi.org/10.1016/J.ENPOL.2018.11.019

Karaboga, D., & Kaya, E. (2018). Adaptive network based fuzzy inference system (ANFIS) training approaches: a comprehensive survey. *Artificial Intelligence Review 2018 52:4*, *52*(4), 2263–2293. https://doi.org/10.1007/S10462-017-9610-2

Kshirsagar, A. (2018). Bio-remediation: Use of nature in a technical way to fight pollution in the long run. *ResearchGate*. https://doi.org/10.13140/RG.2.2.26906.70088

Kshirsagar, A., & Shah, M. (2021). Anatomized study of security solutions for multimedia: deep learning-enabled authentication, cryptography and information hiding. *Advanced Security Solutions for Multimedia*. https://doi.org/10.1088/978-0-7503-3735-9CH7

Kumar, S., & Barua, M. K. (2021). Exploring and measure the performance of the Indian petroleum supply chain. *International Journal of Productivity and Performance Management* (ahead-of-print). https://doi.org/10.1108/IJPPM-12-2020-0640

Liu, W., Liu, W. D., & Gu, J. (2020). Forecasting oil production using ensemble empirical model decomposition based long short-term memory neural network. *Journal of Petroleum Science and Engineering*, *189*(January), 107013. https://doi.org/10.1016/j.petrol.2020.107013

Naderi, M., Khamehchi, E., & Karimi, B. (2019). Novel statistical forecasting models for crude oil price, gas price, and interest rate based on meta-heuristic bat algorithm. *Journal of Petroleum Science and Engineering*, *172*, 13–22. https://doi.org/10.1016/J.PETROL.2018.09.031

Negash, B. M., & Yaw, A. D. (2020). Artificial neural network based production forecasting for a hydrocarbon reservoir under water injection. *Petroleum Exploration and Development*, *47*(2), 383–392. https://doi.org/10.1016/S1876-3804(20)60055-6

Okey Onoh, J., & Peter Eze, G. (2019). Stock market performance of firms in the Nigerian petroleum sector using the ARIMA model approach. *World Journal of Finance and Investment Research*, *4*(1), 2550–7125. www.iiardpub.org

Parth Manoj, P., & Pravinchandra Shah, A. (2014). Fuzzy logic methodology for short term load forecasting. *IJRET: International Journal of Research in Engineering and Technology*, *3*(4), 2321–7308. http://www.ijret.org

Pattanayak, R. M., Behera, H. S., & Panigrahi, S. (2020). A multi-step-ahead fuzzy time series forecasting by using hybrid chemical reaction optimization with pi-sigma higher-order neural network. *Advances in Intelligent Systems and Computing*, *999*, 1029–1041. https://doi.org/10.1007/978-981-13-9042-5_88

Peng, Q., Wen, F., & Gong, X. (2021). Time-dependent intrinsic correlation analysis of crude oil and the US dollar based on CEEMDAN. *International Journal of Finance and Economics*, *26*(1), 834–848. https://doi.org/10.1002/IJFE.1823

Pollo, B. J., Alexandrino, G. L., Augusto, F., & Hantao, L. W. (2018). The impact of comprehensive two-dimensional gas chromatography on oil & gas analysis: Recent advances and applications in petroleum industry. *TrAC Trends in Analytical Chemistry*, *105*, 202–217. https://doi.org/10.1016/J.TRAC.2018.05.007

Qanbari, M., Javadi, S., & Sabbaghi-Nadooshan, R. (2013). The forecasting of Iran natural gas consumption based on neural-fuzzy system until 2020. *International Journal of Smart Electrical Engineering*, *2*(3), 181–184. http://ijsee.iauctb.ac.ir/article_510111.html

Ruiz, M., Alzate-Espinosa, G., Obando, A., & Alvarez, H. (2019). Combined artificial intelligence modeling for production forecast in an oil field. *CTyF - Ciencia, Tecnologia y Futuro, 9*(1), 27–35. https://doi.org/10.29047/01225383.149

Sagheer, A., & Kotb, M. (2019). Time series forecasting of petroleum production using deep LSTM recurrent networks. *Neurocomputing, 323*, 203–213. https://doi.org/10.1016/j.neucom.2018.09.082

Shirvani, A., & Volchenkov, D. (2019). A regulated market under sanctions: On tail dependence between oil, gold, and Tehran stock exchange index. *Journal of Vibration Testing and System Dynamics, 3*(3), 297–311. https://doi.org/10.5890/jvtsd.2019.09.004

Sinha, U., Dindoruk, B., & Soliman, M. (2020). Machine learning augmented dead oil viscosity model for all oil types. *Journal of Petroleum Science and Engineering, 195*, 107603. https://doi.org/10.1016/J.PETROL.2020.107603

Solanki, P., Baldaniya, D., Jogani, D., Chaudhary, B., Shah, M., & Kshirsagar, A. (2021). Artificial intelligence: New age of transformation in petroleum upstream. *Petroleum Research*. https://doi.org/10.1016/J.PTLRS.2021.07.002

Solarin, S. A., Gil-Alana, L. A., & Lafuente, C. (2020). An investigation of long range reliance on shale oil and shale gas production in the U.S. market. *Energy, 195*, 116933. https://doi.org/10.1016/J.ENERGY.2020.116933

Song, M., & Zhou, X. (2019). A casing damage prediction method based on principal component analysis and gradient boosting decision tree algorithm. *SPE Middle East Oil and Gas Show and Conference, MEOS, Proceedings,* 2019. https://doi.org/10.2118/194956-MS

Syed, F. I., Alnaqbi, S., Muther, T., Dahaghi, A. K., & Negahban, S. (2021). Smart shale gas production performance analysis using machine learning applications. *Petroleum Research*. https://doi.org/10.1016/j.ptlrs.2021.06.003

Vilela, M., Oluyemi, G., & Petrovski, A. (2018). Fuzzy data analysis methodology for the assessment of value of information in the oil and gas industry. *IEEE International Conference on Fuzzy Systems, 2018-July*. https://doi.org/10.1109/FUZZ-IEEE.2018.8491628

Wang, J., Niu, T., Du, P., & Yang, W. (2020). Ensemble probabilistic prediction approach for modeling uncertainty in crude oil price. *Applied Soft Computing, 95*, 106509. https://doi.org/10.1016/J.ASOC.2020.106509

Wulandari, R., Farikhin, Surarso, B., & Irawanto, B. (2020). First-order fuzzy time series based on frequency density partitioning for forecasting production of petroleum. *IOP Conference Series: Materials Science and Engineering, 846*(1), 012063. https://doi.org/10.1088/1757-899X/846/1/012063

Yildirim, S., Jothimani, D., Kavaklioglu, C., & Başar, A. (2019). Classification of »hot news» for financial forecast using NLP techniques. *Proceedings -2018 IEEE International Conference on Big Data, Big Data 2018, Xml*, 4719–4722. https://doi.org/10.1109/BigData.2018.862190

Chapter 9

Future of oil and gas (petroleum) industry with AI

9.1 INTRODUCTION

Artificial intelligence (AI) is a machine capable of performing tasks requiring natural intelligence to think like humans or animals (Kshirsagar, 2018). Speech recognition, learning, and problem-solving are some examples of AI. Several challenges and issues are faced by the oil and gas (O&G) industries in handling data and processing (Balaji et al., 2018). In various techniques and processes, a vast amount of information will be generated. For improving the O&G industry performances, this database should be technically and statistically appropriately analyzed. The data analysis and interpreting process can be done using the various types of AI. In the O&G sector, machine learning (ML) is being used to tackle challenges.

AI is rapidly dominating a wide range of industries, posing tremendous opportunities for development and innovation. AI's primary goal is to increase efficiency. AI has already prompted significant modifications and changed competitive healthcare transportation, retail, media, and finance norms (Kshirsagar & Shah, 2021). Companies in these areas produce value by utilizing AI solutions rather than traditional and human-centered business procedures. The value generation process is driven by advanced algorithms instructed on high and valuable data and are constantly fed new data (Azzedin & Ghaleb, 2019). Petroleum businesses are more likely to embrace the latest tech rather than do new things and change their company model.

Moreover, AI benefits firms from many industries, not just those that are digitally proficient. Although companies in the O&G, construction, and mining industries are late adopters of digitalization, they are becoming increasingly dependent on AI solutions. Although the initial AI applications in these industries were discussed in the 1970s, the business began to search for AI implementation prospects more aggressively a few years ago. It corresponds with the gradual expansion of AI abilities and the organization's transition to the O&G theory, whose central objective is to increase the number using new digital techniques. Large petroleum companies worldwide rely heavily on O&G storage, and it is an essential source of data for assessing a company's capacity to grow. After evaluating and auditing the O&G reserve assets in developed nations such as Europe and America, they must be revealed to society by the Securities and Exchange Commission (SEC) of the United States, which has higher criteria for the management and assessment of SEC O&G reserves. There is still a gap in the management and examination of SEC O&G reserves between China and the rest of

DOI: 10.1201/9781003279532-9

the world, mostly seen in low levels of evaluation and management information and management inefficiency (Panzabekova et al., 2019).

Oil prices have been relatively low since 2014, and the loss of institutional knowledge has forced the petroleum sector in the United States to examine its operations and related expenses (Nazir & Rehman, 2021). Intelligent automation may assist close the knowledge gap by collecting the expertise of knowledgeable professionals before they depart and reduce potential business losses owing to a lack of experience. Moreover, automating repetitive procedures might assist these businesses in better managing their resources and data association, increasing safety, and improving profit and productivity. AI has made significant development in the O&G sector in recent decades, in all of its manifestations, from neural networks to genetic enhancement to fuzzy logic (Solanki et al., 2021). Several essential characteristics of an integrated, intelligent software solution include the capacity to merge soft (intelligent) and hard (statistical) computing and the ability to join some AI approaches. The technologies utilized most often in the O&G business are as follows:

- **Genetic Algorithm (GA)** is the most often used technology in the O&G business, which is motivated by the evolutionary factors of organisms in the native surroundings and comprises a stochastic algorithm (Hossein & Ali, 2020).
- **Fuzzy Logic (FL)** uses a knowledge base (database) and a specialized purpose of discussion to convert sharp (discrete) data as input and predict the proper matching output (Okwu & Nwachukwu, 2019).
- An **Artificial Neural Network (ANN)** comprises many basic processing units, each with its activation state, that interact with one another by transmitting different-weighted signals. The total interaction of the units results in a processed output that is combined with external input (Mohammadzaheri et al., 2019; Rahmanifard & Plaksina, 2019).

Due to the 1973–1983 economic crisis, oil consumption increased consistently until the 2008 crisis (Nyangarika & Tang, 2018). There is a low demand for oil futures, which was prompted by skewed statistics on the labor market status in the United States, which indicates that oil prices will have a little increasing tendency and will be relatively constant from 2019 to 2022.

9.2   AI IN RESERVOIR MANAGEMENT

Duan (2018) examines the distribution of SEC O&G fields and the characteristics of managing objects to study the application of AI in monitoring and treating SEC petroleum reservoirs. The study integrates the fundamental asset data with the management of assessment findings to create a refined management approach. The study also discusses how the management system model was designed using the SOA architecture and AI technologies. The main components of reserves management of petroleum energy firms at the moment are improving the information level of reserves assessment management and responding to the needs of the current big data era (Shadravan et al., 2017). Hence it has been discovered that the strategy can efficiently display important data, generate statistics, and manage data. The system can dynamically adapt to the requirement for fundamental data resources in light of the new SEC O&G

reserve standards. Thus, AI facilitates unified and efficient management and provides a strong basis for PetroChina to start on a strategic plan for independent assessment/evaluation in the future, and effectively support and aid management decision-making (Mohaghegh, 2020).

AI, without a doubt today's most significant common mechanism, is rapidly launching industries, posing tremendous opportunities for development and innovation (Koroteev & Tekic, 2021). Koroteev and Tekic (2021) examine how AI affects the O&G business, a key element of the energy sector. The study focuses on the upstream portion of the O&G industry since it is the most capital-intensive and has the most significant number of uncertainties. The study also outlined three probable scenarios for how AI would grow and affect the O&G sector in the future. In reservoir engineering, the study identifies three critical potentials to use AI:

i.   The first has to do with calculations performed using traditional reservoir modeling software (Hassanvand et al., 2018).
ii. In reservoir engineering, the study identifies three key potentials to use AI.
iii. The method may be the same as for upscaling, with the addition of machine or deep learning to speed up and reduce the bias associated with history matching.

To summarize, it is possible to forecast how AI will grow in the O&G sector over the next 5–20 years based on these three scenarios (Castineira et al., 2018).

On the use of AI methods in reservoir characterization and modeling, there are two different views (Anifowose, 2011). The first view sees AI as technology develops. AI will provide a head start in reservoir characteristics and modeling. The opposing view claims that AI approaches are "black boxes" with ambiguous designs and that their conceptions do not adhere to basic petroleum engineering standards. Abdulmalek et al. (2018) and Anifowose et al. (2017) present AI as a "white box," showing fundamental principles, the architectural composition of some of its approaches, and demonstrating practical different characterizations of reservoirs with varied implementations and modeling problems. Using AI methods, O&G reservoir characteristics like water saturation, porosity, lithofacies permeability, and identification of wellbore stability are predicted. The hybridization and ensemble of various AI methods are also explored. The results of this article give a deeper grasp of AI's fundamental ideas and a solid foundation for ongoing research into AI approaches (Carpenter, 2020). Overall, it will assess successful AI applications in petroleum engineering and improve the essential multimodal cooperation synergy among petroleum engineers, mathematicians, and computer scientists.

The assessment of hydrocarbon reserves is a crucial problem for all O&G firms (Ahmed et al., 2017). Oil recovery factor (RF) estimate may now be accomplished using a variety of methods. The accuracy of these approaches is determined by the availability of data, which is heavily influenced by the reservoir age (Mahmoud et al., 2019). In this work, ten characteristics that are available early are used to estimate AI methods used by four RF. The parameters are the starting stock-tank oil, net pay (adequate reservoir thickness), porosity, asset area (reservoir area), initial water saturation, API gravity, effective permeability, and oil viscosity. RF is a problematic side to solve. In this article, four AI technology/models, including radial basis neural networks, ANNs, adaptive neuro-fuzzy inference systems with support vector machines, and subtractive clustering, were assessed to forecast RF using ten reservoir rock and fluid characteristics that were easily accessible early in the reservoir's duration. Because the predicted RF for the testing dataset has the lowest average absolute percentage error of 7.92% and the highest R2 of 0.94, ANNs seem to be the best AI tool for predicting the RF (38 reservoirs). For the first time, an empirical correlation for RF prediction was developed using an ANN model that could readily be simulated and used to predict the RF. On all error assessment criteria tested in this research, the generated correlation outperformed the reported correlations. Additionally, it had the greatest R2 of 0.94, compared to just 0.55 for the Gulstad correlation, which is presently one of the most accurate available correlations.

9.3   AI IN DRILLING

AI comes in various flavors, from biological efficiency to FL derived from a neural network (Cocchi & Mazzeo, 2020). It has made significant progress in recent years toward wide acceptance in the O&G sector. Thus, as per Cocchi and Mazzeo's (2020) study, recent advances in the O&G upstream segment are becoming apparent, showing that the petroleum industry has realized the tremendous potential offered by intelligent systems. Additionally, the research explains how an interconnected, intelligent software solution must include many critical characteristics, including the capacity to combine hard (statistical) and soft (intelligent) computing, as well as the ability to integrate several AI methods. Additionally, the research highlights the methods that are most often used in the O&G industry, which are as follows: ANN (Abdel Azim, 2020), GA (Liang et al., 2017), and FL (Abdulmalek et al., 2018).

The number of AI applications in specific years are based on drilling operations. Hence it concludes that in 2001 the application of AI in O&G industry is 21% (ANN = 14%, hybrid system = 7%, and FL = 2%), its contribution in 2001 gradually decreases to 6% (hybrid system = 3%, ANN = 2%, and FL = 1%), in 2003 it increases by 15% (hybrid system = 8%, FL = 5%, GA = 4%, and ANN = 4%), and in 2004 its contribution is 23% (hybrid system = 6%, GA = 6%, ANN = 6%, and FL = 5%).

With the introduction of advanced sensors stably installed in the wellbore, massive volumes of data, including vital and essential information, are now accessible (Cocchi & Mazzeo, 2020). According to Cocchi and Mazzeo's (2020) study, operator interaction is needed to control the software that analyzes information in real life to optimize the value of these innovative hardware solutions. Only intelligent systems are capable of bringing real-time assessment and decision-making skills to the new hardware. Additionally, the research discusses AI's use in the petroleum industry. Abugharara et al. (2019) discuss the following important uses of AI in drilling operations using real-world borehole optimizations, in which AI systems are utilized to improve downhole parameter verification and drilling performance.

i.   BHA behavior monitoring
ii. DS vibration monitoring
iii. Frictional drag and load transfer (WOB) monitoring
iv. ROP improvement by managing bit wear
v.   Thrust and drilling torque prediction
vi. Estimation of hole cleaning efficiency

The authors also investigate several organizations predicting the economic benefit of AI in the O&G industry in upcoming years. Hence from his analysis as mentioned above and organization records according to Value Global 4 and Mordor Intelligence 5, the market for AI in O&G will reach $2.85 billion by 2025 and will grow at a compound annual growth rate which is an abbreviation of CAGR at 12.14% until 2027.

The O&G sector faces several operational problems, ranging from disconnected surroundings to regular downtime and machine maintenance issues (Nair, 2020). So, Nair (2020) described his views on the integration of AI with the O&G industry. Compared to other industries, O&G have invested less in ML and AI technology. However, O&G companies have begun to embrace AI as a result of their observation of how the technology has assisted other sectors in reducing operating costs and increasing efficiency (Nunoo, 2018). AI is a wide field, and the research discovered that two main applications inside the O&G sector are ML and data science. Thus, the extract concludes that AI systems are capable of automating and optimizing the first stages of the production and exploration lifecycle, which include drilling, geology, seismology, production, reservoir, and petrophysics. They may help reduce risk, increase productivity, and save operational expenses. Precise targeting and the capacity to precisely locate drilling sites may help maximize the return on the investment for any drilling process. Demand for AI in the O&G sector is projected to grow at a 12.66% compound annual growth rate over the next five years, surpassing $2.8 billion in 2022. These predictions, however, are made prior to the Covid-19 interruption.

Most global O&G sources have already been identified and had been manufacturing for many years, making resource management, exploration, production, and drilling difficult (Yin et al., 2018). And that's why petroleum engineers are attempting to employ modern technologies like ANNs to help make decisions and decrease non-productive costs and time. Applications of ANNs in reservoir management, exploration, production, and drilling were divided into four categories (Osarogiagbon et al., 2020). Many applications from the petroleum engineering literature were compiled. Additionally, a defined approach for applying ANNs to any petroleum applications was given, which was achieved via the use of a flowchart that can be utilized as a practical guideline for applying ANNs to any petroleum applications. The method has been simplified into manageable steps that anybody may follow. The petroleum industry's massive data sets enable improved judgment and forecasting of future repercussions.

9.4 AI IN EXPLORATION

Changes in AI in the O&G industry, a key element of the energy sector, are investigated (Koroteev & Tekic, 2021). Koroteev and Tekic (2021) concentrate on the upstream portion of the O&G industry since it is the most capital-intensive and has the greatest uncertainties. The most current developments in creating AI-established tools and highlight their influence on speeding and not taking a risk in processes in the industry based on a study of AI applicable possibilities and an assessment of existing implementations are summarized. AI methods and algorithms, as well as the function and accessibility of data, are described. We also go through the key non-technical issues preventing the widespread applications of AI in the O&G sector, such as data and people cooperation. Although AI is still a hot item in the O&G sector, several uses are shown their value. Exploration and proactive field improvement should become cheaper and faster due to AI, while long-term production margins should increase (Anifowose et al., 2013; Mohammadpoor & Torabi, 2020). The scalability of AI is currently being tested throughout the whole industry. It covered not just the technical but also the non-technical aspects of scalability in this article. Examining how education, corporate culture, and data accessibility influenced the acceleration and vector of AI adoption in the upstream O&G industry. Based on this study, three probable scenarios for expanding AI inside the O&G industry in the next 5–20 years are developed.

The focus of this special problem is on AI applications in the petroleum sector (Hazbeh et al., 2021). The issue's main objective is to provide the petroleum sectors with various uses based on modern AI models and methodologies that may be used to solve difficulties in the sector. This special issue covers both computational intelligence and conventional symbolic applications. The applications covered in this volume lie under the scope of AI approaches deployed, prompting the development of new methodologies to address both classic and contemporary issues. The modeling is inspired by the mechanics of logical thought and is fundamental for the applications in the papers. As AI has progressed, many developers and researchers have begun to utilize hybrid systems, which combine many AI approaches. Hybridization employs many AI approaches in a particular world organization, each method excelling at what it does. Each of these methods can appropriately handle a piece of the problem while also creating a cohesive answer to the entire problem. Hybrid systems are gaining popularity right now, and they appear to be delivering positive results (Zhang et al., 2018). According to the workshop on the potential of AI technologies in the petroleum sector, soft computing, executive computing, and hybrid computing are the greatest candidates to handle the key challenges in the industry. The development of new computing technology for the oil business is influenced by the human–computer framework, becoming increasingly human-like. To accomplish this, research and development of interaction methods and models, cooperation, and competitive tactics, as well as modeling and standardizing human people procedures in the oil sector, are essential.

The main objective of this study was to develop a valid association for forecasting oil rates in gas-lift wells using AI techniques (Khan et al., 2018). Support vector machines, artificial neuro-fuzzy inference systems, functional networks, and ANNs were all utilized in this research. Additionally, ANN was used to develop a physical equation for forecasting the flow rate of oil. Separator test data from many wells in offshore oil with continuous gas lifts were collected (Sabah et al., 2019). Before the data was fed into the algorithm, it underwent thorough data analysis. As inputs, we utilized just the most easily available surface properties. The absolute average percentage error and regression coefficient were used to conduct the analysis. The newly developed AI model is capable of estimating oil prices with a 98% accuracy, which is extremely efficient, and there are no previous instances of such results. When compared to real separator data, ML algorithms correctly predict the rate of oil in an artificial upgrading in a gas well with a 96%–99% accuracy. The authors propose a novel econometric technique for predicting oil flow rate that may be utilized in any situation where the input values fall within the specified model's range. It does not, however, need programming knowledge or the use of compiled code.

Because O&G firms embrace new technologies faster than they experiment with and alter their business models, the main goal of their AI (and other digitalization) initiatives is to increase efficiency (Koroteev & Tekic, 2021). In reality, this usually implies speeding up procedures and lowering risks. Koroteev and Tekic (2021) explored and demonstrated how AI alters the oil or gas upstream. The study has mostly focused on what de-risking involves in the O&G industry and how AI might assist in this process. Exploration for O&G fields includes steps that result in a three-dimensional map of the deposits and geographical model of an oil or gas reservoir. Petrophysical and geophysical investigations and the processing of data collected during the studies are part of the activities: (1) well logging, (2) lab core interpretation, and (3) reservoir-scale seismic surveys (in certain very specialized situations), technical core analysis are typical geophysical and petrophysical investigations. Thus, in this way, in his perspective, AI-assisted technologies are the apparent method to force and, more importantly, eliminate the subjective component of the concluding process.

AI is a crucial element of industrial improvement and supports developing technologies such as the graphic processing unit, blockchain, Internet of Things, and cloud computing in the next generation of big data and firms (Lu, 2019). Lu (2019) compiles a comprehensive overview of AI and deep learning from 1961 to 2018. The study has concentrated chiefly on what escalation involves in the O&G industry and how AI can assist in this procedure. Through a multi-angle comprehensive examination of AI, the research provides a valuable reference for scholars and employees, from basic steps to real-world applications, from basic methods to business accomplishments, and more. With the numerous difficulties surrounding AI, it is undeniable that AI has evolved into a revolutionary and inventive assistance in a broad range of uses and areas.

9.5   AI IN PRODUCTION

Fossil fuel prices continue to climb, and fossil fuel businesses need to create and strengthen advanced technology to maximize productivity and develop their current capacities (Sircar et al., 2021). Other chemicals, such as fertilizers, pharmaceutical drugs, polymers, solvents, and insecticides, need O&G as a source of energy. Sircar et al. (2021) state that technology has a significant impact on society and industry. Digital transformation is considered the fourth industrial revolution, evidenced by AI, self-driving cars, and robots are examples of technology that blur the borders between the real, technological, and organic domains. ML has shown promise in boosting and augmenting conventional hydraulic engineering techniques across many problems. AI technology has received popularity these days because of its scalability and speed of generalization. The ANN structure is a deep learning concept that learns and knows the concepts of data (Osarogiagbon et al., 2021). Neural networks are used to model the data. It is a collection of ML algorithms. The deep learning methods are used to process the vast amount of information in the O&G industry, which analyzes the massive data and produces the best results. Without the help of human Intervention, the best features will be discovered. Moreover, deep learning algorithms can conduct advanced operations that ML algorithms cannot. Neural networks are used to process inputs. ANNs are a powerful ML tool for solving complex issues. The ANN was used in the O&G industry to handle complex nonlinear problems that a linear connection could not solve.

AI is being used to simplify complicated decision-making methodology in almost every testing market area and the upstream O&G industries without reduction

(Bello et al., 2015). AI is the application of advanced networking technologies and algorithms to solve multidimensional issues as it mimics human intelligence, to allow computers and robots to perform jobs that previously required intensive human brainstorming. Bello et al. (2016) describe that unlike other types of computational automation, AI allows the intended tools to learn via repeated use, allowing the system to improve its computing skills as more data is input into it. In the worldwide petroleum exploration and production business, AI has led to substantial design and calculation improvements. Its uses have only grown with the introduction of recent drilling and production technology. AI tools have been conveyed to bridge the technological gaps preventing automated execution and drilling monitoring. The key completion procedures, reservoir simulation such as reservoir classification and PVT analysis, history matching, drill bits diagnosis, permeability prediction, and porosity prediction, well production optimization, overtime well pressure-drop evaluation, seismic pattern recognition, well performance projection, and rapid and making decisions in costly and critical drilling operations (Eyitayo et al., 2020). The study examines and evaluates how AI approaches have been successfully integrated as the missing piece of the jigsaw in a variety of boreholes, reservoirs, and production sectors. It gives new information on the amount of AI engagement in service operations and industrial application trends.

Subsurface engineers can use a model that can give accurate gas rate predictions for a gas reservoir as a useful tool when addressing well and reservoir optimization techniques (Kalam et al., 2019). This paper provided by Kalam et al. (2019) describes AI models for estimating gas rates in a ten-well oil field. The goal is to create an uncomplicated relationship to use and implement while giving reliable results worldwide. A variety of ML tools are used. Adaptive neuro-fuzzy inference systems, ANNs, functional networks are among them. The model was built using production data from a dry gas field named X. Data cleaning and extraction procedures were used to ensure that the proposed model's input parameters were relevant and accurate. If these stages are skipped, plenty of correlation, that is garbage-in garbage-out, would emerge. This resulted in a few fundamental well-head characteristics found in every normal well and directly influenced the output rate of production. The gas rate is the goal parameter for model training. The coefficient of determination and an average absolute percentage error were used to evaluate the AI models under consideration. According to the comparison analysis, the intelligent model can forecast gas rates in condensate wells with an accuracy of over 90%. There have not been any earlier examples with such high precision. When compared to the other adaptive algorithms employed in this study, ANN outperforms them all (Butt et al., 2020). The potential of Industrial Revolution 4.0 for the Pakistan O&G sector is explored in this study. Well test and multiphase flow meters findings may be validated using data-driven AI models.

Since 2014, oil prices have been relatively stable (Blancett et al., 2020). The loss of institutional expertise (called the Great Crew Change) has compelled the United States' petroleum sector to examine its operations and related expenses. Thus, according to Blancett et al. (2020), intelligent automation can help close the gap in knowledge by collecting the expertise of experienced personnel before their departure and minimizing economic opportunity losses owing to a lack of experience. Additionally, automating repetitive procedures enables these businesses to effectively manage their personnel and data, promote safety and cooperation, and boost production and profitability. Thus, their study analyzes the possibilities for automation in every step of the supply chain, from upstream exploration and production through midstream transportation and distribution, downstream refining, and lastly, the wheels at retail gasoline stations. The article does this by defining frameworks for discovering and prioritizing possible automation use cases. Additionally, it specifies nine fundamental characteristics that enable a company to leverage automation to operate its business more efficiently. Automation allows the collection and analysis of huge volumes of data relatively fast. Meanwhile, when obtaining data, the workforce can limit or even eliminate their exposure to dangerous situations (Noshi & Schubert, 2018). Among the other benefits of automation are the elimination of errors and the ability to make better-informed decisions due to real-time data and analysis. Automation integration done correctly and intelligently results in cost and labor savings.

Oil and gas remain the most lucrative commodity markets on the power side, despite the growing adoption of renewable energy alternatives (Wu et al., 2019). Nevertheless, many people are concerned about the enormous environmental impact of fewer energy sources in this day and period of global warming. O&G businesses are very interested in advanced technologies since they are constantly striving to improve and optimize the demand and supply of these resources to go forward of the curve. Wu et al. (2019) emphasize that O&G development is becoming even more reflective of a manufacturing approach, where specific margin enhancement initiatives are the key differentiating factor among peers. There are numerous AI and ML applications in the O&G industry, and a few of them might have a significant influence in a sector that is trying to modernize. ML, predictive modeling, and big data skills in upstream O&G operations may save $50 billion in costs. Most O&G executives have the skills and resources to invest significant human and financial capital in the current AI revolution (Table 9.1).

Table 9.1 Comparative analysis of the studies

Ref.	Year	Contributions	Results
Koroteev and Tekic (2021)	2021	Because O&G companies adopt new technology at a quicker rate than they experiment with and change their business models, the primary objective of their AI projects is to boost efficiency. In practice, this generally entails expediting operations and reducing risks. Discuss the major non-technical obstacles to the widespread use of AI in the O&G sector, including people and data kinds of cooperation. We also lay out three different scenarios for how AI technology will develop in the O&G industry, and what effect it will have on the future generation.	Even if AI is still a new trend in the O&G industry, applications are always valuable. How education, corporate culture, and data accessibility influenced the direction, also the speed of AI adoption in the O&G upstream are investigated. We develop three probable scenarios for how AI may expand inside the O&G sector in the further 5–20 years based on this study.

(Continued)

Table 9.1 (Continued) Comparative analysis of the studies

Ref.	Year	Contributions	Results
Cocchi and Mazzeo (2020)	2020	The recent advancements in the sector of O&G upstream have become noticeable, indicating that the petroleum sector has recognized the enormous skill given by intelligent systems. The most often utilized approaches in the O&G sector, which are ANN, FL, and GA, are discussed. The number of AI applications in specific years based on drilling operations is depicted.	It concludes that in 2001 the application of AI in O&G industry is 21% (ANN = 14%, hybrid system = 7%, and FL = 2%), its contribution in 2001 gradually decreases to 6% (hybrid system = 3%, ANN = 2%, and FL = 1%), in 2003 it increases by 15% (hybrid system = 8%, FL = 5%, GA = 4%, and ANN = 4%), and in 2004 its contribution is 23% (hybrid system = 6%, GA = 6%, ANN = 6%, and FL = 5%).
Blancett et al. (2020)	2020	Their study analyzes the possibilities for automation in every step of the supply chain, from upstream exploration and production through midstream transportation and distribution, downstream refining, and lastly to the wheels at retail gasoline stations. The article does this by defining frameworks for discovering and prioritizing possible automation use cases. It specifies nine fundamental characteristics that enable a company to leverage automation to operate its business more efficiently. Automation allows the collection and analysis of huge volumes of data relatively fast. Meanwhile, when obtaining data, the workforce can limit or even eliminate their exposure to dangerous situations.	This helps to close the gap in knowledge by collecting the expertise of experienced personnel prior to their departure and minimizing economic opportunity losses owing to a lack of experience. Additionally, automating repetitive procedures enables these businesses to effectively manage their personnel and data, promote safety and cooperation, and boost production and profitability.
Hazbeh et al. (2021)	2021	The focus of this special problem is on AI applications in the petroleum sector. The problem's main objective is to provide the petroleum sector with a variety of applications based on high methodologies that may be used to solve difficulties in the sector. This special issue covers both conventional symbolic and computational intelligence applications. The innovative idea of new computing technology for the oil business is influenced by the human–computer framework, which is becoming increasingly human-like.	The applications covered in this volume lie under the scope of AI approaches that have been deployed, prompting the development of new methodologies to address both classic and contemporary issues. As AI has progressed, many developers and researchers have begun to utilize hybrid systems, which combine many AI approaches. Research and development of interaction methods and models, cooperation, competitive tactics, and standardization and modeling of human people operations in the oil sector are essential.

(Continued)

Table 9.1 (Continued) Comparative analysis of the studies

Ref.	Year	Contributions	Results
Khan et al. (2020)	2020	The main objective of this study was to develop a valid association for forecasting oil rates in gas-lift wells using AI techniques. In this research, ANNs, artificial neuro-fuzzy inference systems, functional networks, and support vector machines were utilized as AI methods. Additionally, ANN was used to develop a physical equation for estimating the rate of oil flow. The objective is to create a basic model with the fewest possible parameter uncertainty.	The newly created AI model can estimate rates of oil with a 98% precision, which is very efficient, and no earlier examples of such findings have been recorded. ML algorithms have been found to accurately forecast the rate of oil with a 96%–99% accuracy when results were compared to real separator readings. A novel theoretical model for predicting flow rate of oil is provided, which may be utilized in any situation with input parameters within the established model's range, without requiring coding skills or the use of complicated software.
Duan (2018)	2018	The SEC organization's O&G reserves and the characteristics of administration value are investigated to see if AI can be used to analyze and manage SEC O&G reserves. Then, in favor of trying an improved management model, the fundamental knowledge about assets is combined with the management of assessment outcomes. The management system model was designed using the SOA architecture and AI technologies are discussed.	The system can dynamically adapt to the requirement for fundamental data resources in light of the new SEC O&G reserve standards. Thus, AI facilitates unified and efficient management and provides a strong basis for PetroChina to start on a strategic plan for independent assessment/ evaluation in the future and effectively support and aid management decision-making.

REFERENCES

Abdel Azim, R. (2020). Application of artificial neural network in optimizing the drilling rate of penetration of western desert Egyptian wells. *SN Applied Sciences*, 2(7). https://doi.org/10.1007/s42452-020-2993-8

Abdulmalek, A. S., Elkatatny, S., Abdulraheem, A., Mahmoud, M., Abdulwahab, Z. A., & Mohamed, I. M. (2018). Pore pressure prediction while drilling using fuzzy logic. In *Society of Petroleum Engineers: SPE Kingdom of Saudi Arabia Annual Technical Symposium and Exhibition 2018*, SATS 2018. https://doi.org/10.2118/192318-MS

Abugharara, A. N., Molgaard, J., Hurich, C. A., & Butt, S. D. (2019). Study of the Influence of Controlled Axial Oscillations of pVARD on Generating Downhole Dynamic WOB and Improving Coring and Drilling Performance in Shale. *Proceedings of the International Conference on Offshore Mechanics and Arctic Engineering–OMAE*, 8. https://doi.org/10.1115/OMAE2019-96189

Ahmed, A. A., Elkatatny, S., Abdulraheem, A., & Mahmoud, M. (2017). Application of artificial intelligence techniques in estimating oil recovery factor for water derive sandy reservoirs. *Society of Petroleum Engineers: SPE Kuwait Oil and Gas Show and Conference 2017*. https://doi.org/10.2118/187621-MS

Anifowose, F. A. (2011). Artificial intelligence application in reservoir characterization and modeling: whitening the black box. In *Society of Petroleum Engineers: Saudi Arabia Section Young Professionals Technical Symposium 2011*, 45–57. https://doi.org/10.2118/155413-MS

Anifowose, F. A., Abdulraheem, A., Al-Shuhail, A., & Schmitt, D. P. (2013). Improved permeability prediction from seismic and log data using artificial intelligence techniques. *SPE Middle East Oil and Gas Show and Conference, MEOS, Proceedings, 3*, 2190–2196. https://doi.org/10.2118/164465-ms

Anifowose, F. A., Labadin, J., & Abdulraheem, A. (2017). Ensemble machine learning: An untapped modeling paradigm for petroleum reservoir characterization. *Journal of Petroleum Science and Engineering, 151*, 480–487. https://doi.org/10.1016/j.petrol.2017.01.024

Azzedin, F., & Ghaleb, M. (2019). Towards an architecture for handling big data in oil and gas industries: Service-oriented approach. *International Journal of Advanced Computer Science and Applications, 10*(2), 554–562. https://doi.org/10.14569/ijacsa.2019.0100269

Balaji, K., Rabiei, M., Suicmez, V., Canbaz, C. H., Agharzeyva, Z., Tek, S., Bulut, U., & Temizel, C. (2018). Status of data-driven methods and their applications in oil and gas industry. *Society of Petroleum Engineers: SPE Europec Featured at 80th EAGE Conference and Exhibition 2018*. https://doi.org/10.2118/190812-MS

Bello, O, Holzmann, J, Yaqoob, T., & Teodoriu, E. (2015). Application of artificial intelligence methods in drilling system design and operations: a review of the state of the art. *Journal of Artificial Intelligence and Soft Computing Research*. https://scholar.google.com/scholar?hl=en&as_sdt=0%2C5&q=Opeyemi%2C+B.%2C+Javier%2C+H.%2C+Tanveer%2C+Y.%2C+Catalin%2C+T.%2C+2015.+Application+of+artificial+intelligence+method+in+drilling+system+design+and+operation%3A+a+review+of+the+state+of+the+art%285%29.+121-139&btnG=

Bello, O., Teodoriu, C., Yaqoob, T., Oppelt, J., Holzmann, J., & Obiwanne, A. (2016, August 2). Application of artificial intelligence techniques in drilling system design and operations: A state of the art review and future research pathways. *Society of Petroleum Engineers: SPE Nigeria Annual International Conference and Exhibition*. https://doi.org/10.2118/184320-ms

Blancett, J., Pocker, S., & Ranjan, A. (2020). Digital operations automating the petroleum industry, from wells to wheels. *Technology Solutions, Cognizant*.

Butt, R., Siddiqui, H., Soomro, R. A., & Asad, M. M. (2020). Integration of Industrial Revolution 4.0 and IOTs in academia: A state-of-the-art review on the concept of Education 4.0 in Pakistan. *Interactive Technology and Smart Education, 17*(4), 337–354. https://doi.org/10.1108/ITSE-02-2020-0022

Carpenter, C. (2020). AI-based decline-curve analysis manages reservoir performance. *Journal of Petroleum Technology, 72*(09), 58–59. https://doi.org/10.2118/0920-0058-JPT

Castineira, D., Zhai, X., Darabi, H., Valle, M., Maqui, A., Shahvali, M., & Yunuskhojayev, A. (2018, December 10). Augmented AI solutions for heavy oil reservoirs: Innovative workflows that build from smart analytics, machine learning and expert-based systems. *Society of Petroleum Engineers: SPE International Heavy Oil Conference and Exhibition 2018*, HOCE 2018. https://doi.org/10.2118/193650-MS

Cocchi, M., & Mazzeo, R. L. (2020). *Current trends in Artificial Intelligence (AI) Application to Oil and Gas Industry*. https://www.degruyter.com/downloadpdf/j/jaiscr.2015.5.issue-2/jaiscr-2015-0024/jaiscr-2015-0024.pdf

Duan, X. (2018). Application of artificial intelligence in evaluation and management of SEC oil and gas reserves. *Chemical Engineering Transactions, 71*, 925–930. https://doi.org/10.3303/CET1871155

Eyitayo, S. I., Ekundayo, J. M., & Mumuney, E. O. (2020). Prediction of reservoir saturation pressure and reservoir type in a niger delta field using supervised machine learning ML algorithms. *Society of Petroleum Engineers: SPE Nigeria Annual International Conference and Exhibition 2020*, NAIC 2020. https://doi.org/10.2118/203697-MS

Hassanvand, M., Moradi, S., Fattahi, M., Zargar, G., & Kamari, M. (2018). Estimation of rock uniaxial compressive strength for an Iranian carbonate oil reservoir: Modeling vs. artificial neural network application. *Petroleum Research*, *3*(4), 336–345. https://doi.org/10.1016/J. PTLRS.2018.08.004

Hazbeh, O., Ahmadi Alvar, M., Khezerloo-ye Aghdam, S., Ghorbani, H., Mohamadian, N., & Moghadasi, J. (2021). Hybrid computing models to predict oil formation volume factor using multilayer perceptron algorithm. *Journal of Petroleum and Mining Engineering*, *23*(1), 17–30. https://doi.org/10.21608/JPME.2021.52149.1062

Hossein, M., & Ali, R. M. (2020). A nonlinear approach for predicting pore pressure using genetic algorithm in one of the Iranian petroleum carbonate reservoirs. *Arabian Journal of Geosciences*, *13*(14), 1–13. https://doi.org/10.1007/S12517-020-05692-1

Kalam, S., Khan, M. R., Tariq, Z., Siddique, F. A., Abdulraheem, A., & Khan, R. A. (2019). A novel correlation to predict gas flow rates utilizing artificial intelligence: An industrial 4.0 approach. *Society of Petroleum Engineers: SPE/PAPG Pakistan Section Annual Technical Symposium and Exhibition 2019*, PATS 2019. https://doi.org/10.2118/201170-MS

Khan, M. R., Tariq, Z., & Abdulraheem, A. (2018). Machine learning derived correlation to determine water saturation in complex lithologies. *Society of Petroleum Engineers: SPE Kingdom of Saudi Arabia Annual Technical Symposium and Exhibition 2018*, SATS 2018. https://doi.org/10.2118/192307-MS

Khan, M. R., Tariq, Z., & Abdulraheem, A. (2020). Application of artificial intelligence to estimate oil flow rate in gas-lift wells. *Natural Resources Research*, *29*(6), 4017–4029. https://doi.org/10.1007/S11053-020-09675-7

Koroteev, D., & Tekic, Z. (2021). Artificial intelligence in oil and gas upstream: Trends, challenges, and scenarios for the future. *Energy and AI*, *3*, 100041. https://doi.org/10.1016/j.egyai.2020.100041

Kshirsagar, A. (2018). Bio-remediation: Use of nature in a technical way to fight pollution in the long run. *ResearchGate*. https://doi.org/10.13140/RG.2.2.26906.70088

Kshirsagar, A., & Shah, M. (2021). Anatomized study of security solutions for multimedia: deep learning-enabled authentication, cryptography and information hiding. *Advanced Security Solutions for Multimedia*. https://doi.org/10.1088/978-0-7503-3735-9CH7

Liang, H., Zou, J., & Liang, W. (2017). An early intelligent diagnosis model for drilling overflow based on GA–BP algorithm. *Cluster Computing*, *22*(5), 10649–10668. https://doi.org/10.1007/S10586-017-1152-5

Lu, Y. (2019). Artificial intelligence: a survey on evolution, models, applications and future trends. *Journal of Management Analytics, 6*(1), 1–29. https://doi.org/10.1080/23270012.2019.1570365

Mahmoud, A. A., Elkatatny, S., Chen, W., & Abdulraheem, A. (2019). Estimation of oil recovery factor for water drive sandy reservoirs through applications of artificial intelligence. *Energies*, *12*(19), 3671. https://doi.org/10.3390/EN12193671

Mohaghegh, S. D. (2020). Subsurface analytics: Contribution of artificial intelligence and machine learning to reservoir engineering, reservoir modeling, and reservoir management. *Petroleum Exploration and Development*, *47*(2), 225–228. https://doi.org/10.1016/S1876-3804(20)60041-6

Mohammadpoor, M., & Torabi, F. (2020). Big Data analytics in oil and gas industry: An emerging trend. *Petroleum*, *6*(4), 321–328. https://doi.org/10.1016/j.petlm.2018.11.001

Mohammadzaheri, M., Tafreshi, R., Khan, Z., Ghodsi, M., Franchek, M., & Grigoriadis, K. (2019). Modelling of petroleum multiphase flow in electrical submersible pumps with shallow artificial neural networks. *Ships and Offshore Structures*, *15*(2), 174–183. https://doi.org/10.1080/17445302.2019.1605959

Nair, S. (2020). *The Growing Importance of AI in the Oil and Gas Industry | GEP*. GEP. https://www.gep.com/blog/mind/the-growing-importance-of-ai-in-the-oil-and-gas-industry

Nazir, S., & Rehman, S. U. L. (2021). Trends in crude oil price and its economic impact. *South Asian Journal of Marketing & Management Research*, *11*(5), 2249–2877. https://doi.org/10.5958/2249-877X.2021.00038.2

Noshi, C. I., & Schubert, J. J. (2018). The role of machine learning in drilling operations; a review. *SPE Eastern Regional Meeting*. https://doi.org/10.2118/191823-18ERM-MS

Nunoo, N. A. (2018). Guest editorial: How artificial intelligence will benefit drilling. *Journal of Petroleum Technology*, *70*(05), 14–15. https://doi.org/10.2118/0518-0014-jpt

Nyangarika, A. M., & Tang, B. (2018). Oil price factors: Forecasting on the base of modified ARIMA model. *IOP Conference Series: Earth and Environmental Science*, *192*(1), 012058. https://doi.org/10.1088/1755-1315/192/1/012058

Okwu, M. O., & Nwachukwu, A. N. (2019). A review of fuzzy logic applications in petroleum exploration, production and distribution operations. *Journal of Petroleum Exploration and Production Technology*, *9*(2), 1555–1568). https://doi.org/10.1007/s13202-018-0560-2

Osarogiagbon, A., Muojeke, S., Venkatesan, R., Khan, F., & Gillard, P. (2020). A new methodology for kick detection during petroleum drilling using long short-term memory recurrent neural network. *Process Safety and Environmental Protection*, *142*, 126–137. https://doi.org/10.1016/J.PSEP.2020.05.046

Osarogiagbon, A. U., Khan, F., Venkatesan, R., & Gillard, P. (2021). Review and analysis of supervised machine learning algorithms for hazardous events in drilling operations. *Process Safety and Environmental Protection*, *147*, 367–384. https://doi.org/10.1016/j.psep.2020.09.038

Panzabekova, A., Ha, N. A., & Suleimenova, A. (2019). Influence of the USA and China on the transformation of the world oil market. *Economics: Strategy and Practice*, *3*(14). https://orcid.org//0000-0002-6389-9637

Rahmanifard, H., & Plaksina, T. (2019). Application of artificial intelligence techniques in the petroleum industry: a review. In *Artificial Intelligence Review*, *52*(4), 2295–2318. https://doi.org/10.1007/s10462-018-9612-8

Sabah, M., Talebkeikhah, M., Agin, F., Talebkeikhah, F., & Hasheminasab, E. (2019). Application of decision tree, artificial neural networks, and adaptive neuro-fuzzy inference system on predicting lost circulation: A case study from Marun oil field. *Journal of Petroleum Science and Engineering*, *177*, 236–249. https://doi.org/10.1016/J.PETROL.2019.02.045

Shadravan, A., Tarrahi, M., & Amani, M. (2017). Intelligent tool to design drilling, spacer, cement slurry, and fracturing fluids by use of machine-learning algorithms. *SPE Drilling and Completion*, *32*(2), 131–140. https://doi.org/10.2118/175238-pa

Sircar, A., Yadav, K., Rayavarapu, K., Bist, N., & Oza, H. (2021). Application of machine learning and artificial intelligence in oil and gas industry. *Petroleum Research*. https://doi.org/10.1016/J.PTLRS.2021.05.009

Solanki, P., Baldaniya, D., Jogani, D., Chaudhary, B., Shah, M., & Kshirsagar, A. (2021). Artificial intelligence: New age of transformation in petroleum upstream. *Petroleum Research*. https://doi.org/10.1016/J.PTLRS.2021.07.002

Wu, T., Mao, Y., & Zhao, G. (2019). *A Model Designed for HSE Big Data Analysis in Petroleum Industry*. https://doi.org/10.2523/iptc-19508-ms

Yin, Q., YANG, J., ZHOU, B., JIANG, M., CHEN, X., FU, C., YAN, L., LI, L., LI, Y., & LIU, Z. (2018). Improve the drilling operations efficiency by the big data mining of real-time logging. *Proceedings of the SPE/IADC Middle East Drilling Technology Conference and Exhibition, 2018*. https://doi.org/10.2118/189330-MS

Zhang, B., Yang, J., Sun, T., Ye, J., Liu, Z., Wu, Y., Xie, R., & Zhou, B. (2018). Prediction model of shallow geological hazards in deepwater drilling based on a hybrid computational approach. *Paper Presented at the 52nd U.S. Rock Mechanics/Geomechanics Symposium*.

Index

Printed in the United States
by Baker & Taylor Publisher Services